Lecture Notes in Computer Science 13609

More information about this series at https://link.springer.com/bookseries/558

Anirban Mukhopadhyay · Ilkay Oksuz ·
Sandy Engelhardt · Dajiang Zhu ·
Yixuan Yuan (Eds.)

Deep Generative Models

Second MICCAI Workshop, DGM4MICCAI 2022
Held in Conjunction with MICCAI 2022
Singapore, September 22, 2022
Proceedings

Springer

Editors
Anirban Mukhopadhyay
TU Darmstadt
Darmstadt, Germany

Ilkay Oksuz
Istanbul Technical University
Istanbul, Turkey

Sandy Engelhardt
University Hospital Heidelberg
Heidelberg, Germany

Dajiang Zhu
The University of Texas at Arlington
Arlington, TX, USA

Yixuan Yuan
University of Hong Kong
Hong Kong, Hong Kong

ISSN 0302-9743 ISSN 1611-3349 (electronic)
Lecture Notes in Computer Science
ISBN 978-3-031-18575-5 ISBN 978-3-031-18576-2 (eBook)
https://doi.org/10.1007/978-3-031-18576-2

This Springer imprint is published by the registered company Springer Nature Switzerland AG
The registered company address is: Gewerbestrasse 11, 6330 Cham, Switzerland

DGM4MICCAI 2022 Preface

It was our genuine honor and great pleasure to hold the second Workshop on Deep Generative Models for Medical Image Computing and Computer Assisted Intervention (DGM4MICCAI 2022), a satellite event at the 25th International Conference on Medical Image Computing and Computer Assisted Intervention (MICCAI 2022).

DGM4MICCAI 2022 was a single-track, half-day workshop consisting of high-quality, previously unpublished papers, presented orally (in a hybrid format), intended to act as a forum for computer scientists, engineers, clinicians, and industrial practitioners to present their recent algorithmic developments, new results, and promising future directions in deep generative models. Deep generative models, such as generative adversarial networks (GANs) and variational auto-encoders (VAEs), are currently receiving widespread attention from not only the computer vision and machine learning communities but also in the MICCAI community. These models combine advanced deep neural networks with classical density estimation (either explicit or implicit) to achieve state-of-the-art results. As such, DGM4MICCAI 2022 provided an all-round experience for deep discussion, idea exchange, practical understanding, and community building around this popular research direction.

This year's DGM4MICCAI was held on September 22, 2022, in Singapore. There was a very positive response to the call for papers for DGM4MICCAI 2022. We received 15 submissions for the workshop. Each paper was reviewed by at least three reviewers and we ended up with 12 accepted papers for the workshop. The accepted papers present fresh ideas on broad topics ranging from methodology (causal inference, latent interpretation, generative factor analysis, etc.) to applications (mammography, vessel imaging, surgical videos, etc.).

The high quality of the scientific program of DGM4MICCAI 2022 was due first to the authors who submitted excellent contributions and second to the dedicated collaboration of the international Program Committee and the other researchers who reviewed the papers. We would like to thank all the authors for submitting their valuable contributions and for sharing their recent research activities.

We are particularly indebted to the Program Committee members and to all the external reviewers for their precious evaluations, which permitted us to set up this proceedings. We were also very pleased to benefit from the keynote lectures of the invited speakers: Michal Rosen-Zvi, IBM Research, Israel, and Islem Rekin, Istanbul Technical University, Turkey. We would like to express our sincere gratitude to these renowned experts for making the second workshop a successful platform to rally forward the deep generative models research within the MICCAI context.

August 2022

Anirban Mukhopadhyay
Ilkay Oksuz
Sandy Engelhardt
Dajiang Zhu
Yixuan Yuan

Organization

Organizing Committee/Program Committee Chairs

Sandy Engelhardt	University Hospital Heidelberg, Germany
Ilkay Oksuz	Istanbul Technical University, Turkey
Dajiang Zhu	University of Texas at Arlington, USA
Yixuan Yuan	City University of Hong Kong, China
Anirban Mukhopadhyay	Technische Universität Darmstadt, Germany

Program Committee

Li Wang	University of Texas at Arlington, USA
Tong Zhang	Peng Cheng Laboratory, China
Ping Lu	University of Oxford, UK
Roxane Licandro	Medical University of Vienna, Austria
Chen Qin	University of Edinburgh, UK
Veronika Zimmer	Technische Universität München, Germany
Dwarikanath Mahapatra	Inception Institute of Artificial Intelligence, UAE
Michael Sdika	CREATIS, INSA Lyon, France
Jelmer Wolterink	University of Twente, The Netherlands
Alejandro Granados	King's College London, UK
Jinglei Lv	University of Sydney, Australia
Shiba Kuanar	Mayo Clinic, USA
Onat Dalmaz	Bilkent University, Turkey
Yuri Tolkach	University Hospital Cologne, Germany

Student Organizers

Lalith Sharan	University Hospital Heidelberg, Germany
Henry Krumb	Technische Universität Darmstadt, Germany
Moritz Fuchs	Technische Universität Darmstadt, Germany
Amin Ranem	Technische Universität Darmstadt, Germany
Caner Özer	Istanbul Technical University, Turkey
Chen Zhen	City University of Hong Kong, China
Guo Xiaoqing	City University of Hong Kong, China

Additional Reviewers

Chen Chen
David Kügler
Martin Menten
Soumick Chatterjee
Arijit Patra
Isabel Funke
Ouyang Cheng

Contents

Methods

Flow-Based Visual Quality Enhancer for Super-Resolution Magnetic Resonance Spectroscopic Imaging

Siyuan Dong[1(✉)], Gilbert Hangel[2], Eric Z. Chen[3], Shanhui Sun[3], Wolfgang Bogner[2], Georg Widhalm[4], Chenyu You[1], John A. Onofrey[5], Robin de Graaf[5], and James S. Duncan[1,5]

[1] Electrical Engineering, Yale University, New Haven, CT, USA
s.dong@yale.edu

[2] Biomedical Imaging and Image-Guided Therapy, Highfield MR Center, Medical University of Vienna, Vienna, Austria

[3] United Imaging Intelligence, Cambridge, MA, USA

[4] Neurosurgery, Medical University of Vienna, Vienna, Austria

[5] Radiology and Biomedical Imaging, Yale University, New Haven, CT, USA

Abstract. Magnetic Resonance Spectroscopic Imaging (MRSI) is an essential tool for quantifying metabolites in the body, but the low spatial resolution limits its clinical applications. Deep learning-based super-resolution methods provided promising results for improving the spatial resolution of MRSI, but the super-resolved images are often blurry compared to the experimentally-acquired high-resolution images. Attempts have been made with the generative adversarial networks to improve the image visual quality. In this work, we consider another type of generative model, the flow-based model, of which the training is more stable and interpretable compared to the adversarial networks. Specifically, we propose a flow-based enhancer network to improve the visual quality of super-resolution MRSI. Different from previous flow-based models, our enhancer network incorporates anatomical information from additional image modalities (MRI) and uses a learnable base distribution. In addition, we impose a guide loss and a data-consistency loss to encourage the network to generate images with high visual quality while maintaining high fidelity. Experiments on a ^{1}H-MRSI dataset acquired from 25 high-grade glioma patients indicate that our enhancer network outperforms the adversarial networks and the baseline flow-based methods. Our method also allows visual quality adjustment and uncertainty estimation. Our code is available at https://github.com/dsy199610/Flow-Enhancer-SR-MRSI.

Keywords: Super-resolution · Brain MRSI · Normalizing flow

1 Introduction

Magnetic Resonance Spectroscopic Imaging (MRSI) is a technique for measuring metabolite concentrations within the body [6]. Because the metabolic level is a crucial indicator of cell activities, MRSI is becoming a valuable tool for studying

A. Mukhopadhyay et al. (Eds.): DGM4MICCAI 2022, LNCS 13609, pp. 3–13, 2022.
https://doi.org/10.1007/978-3-031-18576-2_1

different diseases such as brain tumors [14] and cancers [5]. However, due to the low concentrations of metabolites, current applications of MRSI are limited to low spatial resolutions. Hence, developing a post-processing algorithm for generating higher resolution MRSI will greatly benefit its clinical applications.

Recent advances in deep learning have provided promising results for super-resolution (SR) MRSI [3,16]. These works trained neural networks to map low-resolution MRSI metabolic maps to higher-resolution ones with a pixelwise mean-squared error (MSE) loss function. However, SR is a one-to-many problem, and training with MSE learns a pixelwise average of all possible solutions [20,24]. This can result in blurry SR images with suboptimal visual quality and a lack of high-frequency details [10]. To approach this issue, a few recent works proposed to add adversarial loss to improve the visual quality [12,13,21], but it is well-known that the generative adversarial networks suffer from training instability and mode collapse [1,23]. The normalizing flow (NF) model is a relatively new class of generative models that learns the target distribution via the maximization of exact log-likelihood, making the training more interpretable and stable [1,8,19,23]. NF learns an invertible mapping from the target image distribution to a simple base distribution during training, so the target images can be generated by sampling from the base distribution during inference. A few recent works applied NF on SR of natural images [22,23] and reconstructions of medical images [7,17], and these flow-based methods can generate images with high visual quality.

In this work, we propose a flow-based enhancer network to recover high-frequency details and improve the visual quality of the blurry SR MRSI images given by the SR networks. Borrowing the idea of sharpness enhancement [10], we regard the visual quality enhancement as a subsequent step of the SR network. Hence, the enhancer network only needs to focus on improving the visual quality, not the entire SR process. To boost the performance, we make several modifications to the existing flow-based SR method [23], including incorporating MRI anatomical information and a learnable base distribution. We also enforce a data-consistency (DC) loss to encourage the enhanced images not to alter the original measurements from the scanner, which is very important for medical images. Experimental results show that our enhancer network successfully improves the visual quality of SR metabolic maps, outperforming the adversarial networks and the baseline flow-based methods. Our method also allows visual quality adjustment and uncertainty estimation within the same network.

2 Methods

2.1 Problem Formulation

Given a low-resolution metabolic map $L \in \mathbb{R}^{n \times n}$, the SR network S learns a mapping $H = S(L)$ such that the super-resolved metabolic map $H \in \mathbb{R}^{N \times N}$ is close to the ground truth high-resolution map $I \in \mathbb{R}^{N \times N}$. However, the SR network S trained with pixelwise loss [3,16] and structural loss [12] could result in blurry H. We develop an enhancer network E to improve the visual quality of H, i.e. $H_{enh} = E(H)$, such that H_{enh} has visual quality comparable to I.

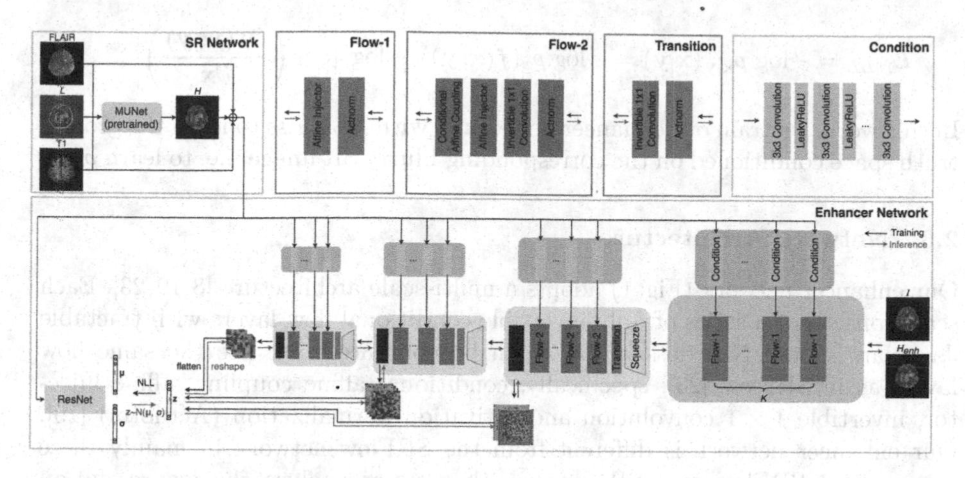

Fig. 1. Proposed flow-based enhancer network for improving the visual quality of SR MRSI. We first use a pretrained Multi-encoder UNet (MUNet) [9,12,13] as the SR Network to obtain a super-resolved image H from the low-resolution image L. The enhancer network takes $\{H, \text{T1}, \text{FLAIR}\}$ as the condition, which is processed with the Condition blocks before feeding into the flow layers. The flow layers Flow-2 and Transition follow the design in SRFlow [23]. We add a special type of flow layer, Flow-1, before any squeeze operation to process the image in its original dimension [25]. During training, the enhancer network transforms the ground truth images I into the Gaussian vectors $\mathbf{z}$ using the NLL loss. During inference, $\mathbf{z}$ is drawn from the Gaussian distribution and inversely passed through the network to obtain H_{enh}. The mean $\boldsymbol{\mu}$ and standard deviation $\boldsymbol{\sigma}$ of the Gaussian distribution is learned from the condition $\{H, \text{T1}, \text{FLAIR}\}$ using a ResNet [15].

2.2 Preliminary: Conditional Normalizing Flow

NF is a family of generative models that constructs a complex distribution from a simple distribution using a flow of invertible transformations [27]. The key idea is to learn a bijective mapping between the target space and a latent space [8]. Given that $\mathbf{x}$ is a sample from the target space with distribution $p_{\mathbf{x}}(\mathbf{x})$, flow-based models usually use an invertible neural network f to transform $\mathbf{x}$ into a latent variable $\mathbf{z}$ with a simple base distribution $p_{\mathbf{z}}(\mathbf{z})$, e.g. Gaussian. Once the transformation is learned, the network f can generate samples from the target distribution $\mathbf{x} = f^{-1}(\mathbf{z})$ by sampling $\mathbf{z} \sim p_{\mathbf{z}}(\mathbf{z})$. This idea can also be used to learn a conditional distribution $p_{\mathbf{x}|\mathbf{y}}(\mathbf{x}|\mathbf{y})$ over two random variables $\mathbf{x}$ and $\mathbf{y}$, also known as the conditional NF [23,31]. According to the change of variable formula, the target distribution $p_{\mathbf{x}|\mathbf{y}}(\mathbf{x}|\mathbf{y})$ can be expressed as

$$p_{\mathbf{x}|\mathbf{y}}(\mathbf{x}|\mathbf{y}) = p_{\mathbf{z}}(f(\mathbf{x};\, \mathbf{y})) \left| \det \left(\frac{\partial f(\mathbf{x};\mathbf{y})}{\partial \mathbf{x}} \right) \right| \tag{1}$$

where $\frac{\partial f(\mathbf{x};\mathbf{y})}{\partial \mathbf{x}}$ is the Jacobian matrix. This expression allows training the network f with the negative log-likelihood (NLL) loss for training samples $(\mathbf{x}, \mathbf{y})$

$$\mathcal{L}_{NLL} = -\log p_{\mathbf{x}|\mathbf{y}}(\mathbf{x}|\mathbf{y}) = -\log p_{\mathbf{z}}(f(\mathbf{x};\mathbf{y})) - \log \left| \det \left(\frac{\partial f(\mathbf{x};\mathbf{y})}{\partial \mathbf{x}} \right) \right|. \tag{2}$$

In this work, we train our enhancer network E with NLL loss to learn the ground truth space conditioned on the corresponding blurry SR image, i.e. to learn $p_{I|H}$.

2.3 Network Architecture

Our enhancer network (Fig. 1) adopts a multi-scale architecture [8,19,23]. Each scale consists of a series of fully invertible conditional flow layers with tractable Jacobian, so the NLL loss in Eq. 2 can be computed. We use the same flow layers as in SRFlow [23], specifically, conditional affine coupling, affine injector, invertible 1×1 convolution and activation normalization (Actnorm) [19]. Our enhancer network is different from the SRFlow network in mainly three aspects: (1) SRFlow is a SR network that super-resolves the low-resolution images, whereas our method enhances the super-resolved images given by any SR network that are blurry; (2) SRFlow processes a single image modality, but our network incorporates information from other modalities (T1 and FLAIR MRI); (3) SRFlow uses a fixed base distribution, standard Gaussian $p_{\mathbf{z}} = \mathcal{N}(0, I)$, but our network adopts a learnable base distribution.

MRI Anatomical Information. Previous works suggested that the multi-parametric MRI contains useful prior information for SR MRSI [12,13]. Therefore, we provide T1 and FLAIR MRI as additional conditions to the enhancer network. Specifically, T1 and FLAIR are re-sampled and concatenated with the super-resolved image H (we denote this as $\{H, \text{T1}, \text{FLAIR}\}$), which are then passed into the Condition blocks of the enhancer network (see Fig. 1).

Learnable Base Distribution. Current flow-based SR models [22,23] assume a 0-mean and unit-norm multivariate Gaussian for the base distribution, whereas such a predetermined base distribution might limit the learning capability of the model. We modify it as a multivariate Gaussian with learnable mean and standard deviation, i.e. $p_{\mathbf{z}} = \mathcal{N}(\boldsymbol{\mu}(c), \boldsymbol{\sigma}(c))$, where the mean and standard deviation vectors are learned from the condition $c = \{H, \text{T1}, \text{FLAIR}\}$ using a ResNet [15]. This "conditional base" was included in the original conditional NF work [31] but was neglected in later applications.

2.4 Loss Function

We apply the NLL loss to learn the conditional distribution of ground truth images I given the SR images H. Therefore, the NLL loss in Eq. 2 becomes

$$\begin{aligned}\mathcal{L}_{NLL}(I) &= -\log p_{\mathbf{z}}(E(I;H)) - \log \left| \det \left(\frac{\partial E(I;H)}{\partial I} \right) \right| \\ &= -\frac{1}{2}\left(\left\| \frac{\mathbf{z} - \boldsymbol{\mu}(c)}{\boldsymbol{\sigma}(c)} \right\|_2^2 + \left\| \log(2\pi\boldsymbol{\sigma}(c)^2) \right\|_1 \right) - \log \left| \det \left(\frac{\partial E(I;H)}{\partial I} \right) \right|.\end{aligned} \tag{3}$$

We define $\mathbf{z} = E(I;H)$ as the training direction and $H_{enh} = E^{-1}(\mathbf{z};H)$ as the inference direction. The second equation holds because $p_{\mathbf{z}} = \mathcal{N}(\boldsymbol{\mu}(c), \boldsymbol{\sigma}(c))$. The log-determinant term can be computed efficiently, because the flow layers are designed to have tractable Jacobian [19].

We add a guide loss in the inference direction to guide the network to learn an enhanced image space centered around the fidelity-oriented SR image [22]

$$\mathcal{L}_{guide}(I, H_{enh}^{\tau=0}) = (1-\alpha)\mathcal{L}_{pixel}(I, H_{enh}^{\tau=0}) + \alpha\mathcal{L}_{structural}(I, H_{enh}^{\tau=0}) \quad (4)$$

where the temperature τ is a scale parameter that controls the variance of the random sample: $H_{enh}^{\tau=\tau_0} = E^{-1}(\mathbf{z};H)$ with $\mathbf{z} \sim \mathcal{N}(\boldsymbol{\mu}(c), \tau_0\boldsymbol{\sigma}(c))$. The pixel loss L_{pixel} defines a pixelwise difference between two images using L1-norm, and the structural loss $L_{structural}$ maximizes the Multiscale Structural Similarity (MS-SSIM) [30] between two images [12,13].

Furthermore, we use a DC loss [4] to encourage that the enhanced image follows the k-space measurement from the scanner

$$\mathcal{L}_{DC}(L, H_{enh}^{\tau=\tilde{\tau}}) = \left\|\mathcal{F}(L) - \mathcal{F}_u(H_{enh}^{\tau=\tilde{\tau}})\right\|_1 \quad (5)$$

where $\mathcal{F}$ is the Fourier transform operator, and $\mathcal{F}_u$ denotes down-sampling after Fourier transform to match the dimension of the low-resolution measurement. $\tilde{\tau}$ is uniformly generated $\tilde{\tau} \sim \mathcal{U}(0,1)$ during training to encourage DC at all temperature levels. DC loss instructs the enhancer network not to modify the scanner measurement, which is very important for reliable clinical diagnosis.

Overall, the enhancer network is trained with:

$$\mathcal{L} = \mathcal{L}_{NLL}(I) + \lambda_1\mathcal{L}_{guide}(I, H_{enh}^{\tau=0}) + \lambda_2\mathcal{L}_{DC}(L, H_{enh}^{\tau=\tilde{\tau}}) \quad (6)$$

3 Experiments

3.1 Data Acquisition and Preprocessing

We acquired 3D ^{1}H-MRSI, T1 and FLAIR from 25 high-grade glioma patients using a Siemens 7T whole-body-MR imager [14]. IRB approval and informed consent from all participants were obtained. MRI images were skull-stripped using FSL v5.0 [28]. The MRSI sequences were acquired using an acquisition delay of 1.3 ms, repetition time of 450 ms and scan duration of 15 min. The nominal resolution is $3.4 \times 3.4 \times 3.4\,\text{mm}^3$, and the matrix size is $64 \times 64 \times 39$. Note that this is a very high resolution for MRSI because of the challenges in acquiring metabolite signals with acceptable SNR. The MRSI spectra were quantified using LCModel v6.3-1 [26] to obtain the 3D metabolic maps. The voxels with insufficient quality (SNR < 2.5 or FWHM > 0.15 ppm) or under strong distortion around the brain periphery were excluded. We focused on 7 metabolites that are major markers of onco-metabolism [14], namely N-acetyl-aspartate (NAA), total creatine (tCr), total choline (tCho), glutamate (Glu), glutamine (Gln), inositol (Ins), and glycine (Gly).

Table 1. Quantitative results. Results are presented in mean ± standard deviation. A lower LPIPS score means better results (↓). The best scores are shown in bold.

Type	Methods	PSNR (↑)	SSIM (↑)	LPIPS (↓)
Fidelity-oriented	MUNet [12]	**29.7 ± 2.5**	**0.933 ± 0.028**	0.0897 ± 0.0462
Visual-oriented	MUNet-cWGAN [12]	28.3 ± 2.6	0.920 ± 0.028	0.0529 ± 0.0349
	SRFlow [23]	27.8 ± 2.5	0.905 ± 0.048	0.0656 ± 0.0516
	Flow Enhancer(ours)	29.0 ± 2.4	0.924 ± 0.029	**0.0519 ± 0.0340**

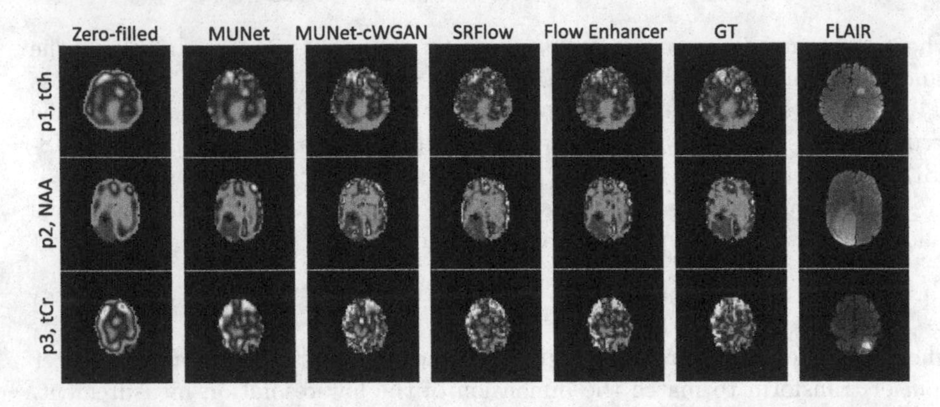

Fig. 2. Qualitative results. From left to right: k-space zero-filled images, MUNet, MUNet-cWGAN, SRFlow, our Flow Enhancer, ground truth (GT) and FLAIR images for anatomical reference. From top to bottom: tCh, NAA and tCr maps from three different patients p1, p2 and p3, respectively. Zoom in to inspect more details.

3.2 Implementation Details

From every 3D MRSI scan, we obtained 9–18 axial slices, and each includes 7 metabolites, summing to 2275 2D metabolic maps. These are regarded as the high-resolution ground truth $I \in \mathbb{R}^{64\times64}$, which were truncated in k-space to obtain low-resolution images $L \in \mathbb{R}^{16\times16}$. Of the 25 patients, we used 15 for training, 5 for validation and 5 for testing. This corresponds to 1246 metabolic maps for training, 483 for validation and 546 for testing. The training dataset was augmented using random rotation and shifting at every training iteration.

The enhancer network operates in 4 scales, each contains $K = 12$ flow steps. The loss weighting parameters are $\alpha = 0.84$ [33], $\lambda_1 = 10$ and $\lambda_2 = 10$. The experiments were implemented in PyTorch v1.1.0 and performed on NVIDIA GTX 1080 and V100 GPUs. The networks were trained with the Adam optimizer [18], initial learning rate of 1×10^{-4}, batch size of 8 and 500 epochs.

3.3 Results

We implemented the enhancer network to improve the visual quality of the SR images given by a Multi-encoder UNet (MUNet) [9,12,13], which was trained

Table 2. Ablation Studies. ✓ or × represents whether a certain design element is present or not. Paired t-test was performed between our method (last row) and the first four rows. Statistically significant differences (p-value < 0.05) are shown with *.

MRI prior	Cond. Base	$\mathcal{L}_{guide}$	$\mathcal{L}_{DC}$	PSNR (↑)	SSIM (↑)	LPIPS (↓)
×	✓	✓	✓	28.9 ± 2.3*	0.920 ± 0.034*	0.0558 ± 0.0367*
✓	×	✓	✓	**29.0 ± 2.4***	**0.924 ± 0.029**	0.0526 ± 0.0337*
✓	✓	×	×	28.3 ± 2.3*	0.918 ± 0.029*	0.0579 ± 0.0398*
✓	✓	✓	×	28.8 ± 2.3*	0.922 ± 0.029*	**0.0513 ± 0.0339***
✓	✓	✓	✓	**29.0 ± 2.4**	**0.924 ± 0.029**	0.0519 ± 0.0340

with a pixelwise plus structural loss. The MUNet uses two encoders for processing anatomical information in T1 and FLIAR respectively. For comparison, we implemented two previous visual-oriented SR methods (these methods were applied to L): (1) MUNet trained with the adversarial loss using conditional Wasserstein generative adversarial networks (MUNet-cWGAN) [12], and (2) the baseline flow-based SR model that does not have our new design elements (e.g. MRI condition, $\mathcal{L}_{DC}$), denoted as SRFlow [23]. We set $\tau = 0.8$ for the flow-based methods as recommended by previous works [22,23]. As for the evaluation metrics, peak signal-to-noise ratio (PSNR) and structural similarity index (SSIM) are the most commonly used metrics for evaluating image SR, but they are ineffective in measuring image visual quality [10]. Previous literature indicates that these fidelity-oriented metrics are often degraded as the visual quality improves [2,23,29]. Here we report a visual-oriented metric, Learned Perceptual Image Patch Similarity (LPIPS), which measures the high-level similarity between two images using a pretrained deep network (AlexNet) and correlates well with human perceptual judgment [10,13,32]. Table 1 shows that although MUNet achieves high PSNR and SSIM scores, its LPIPS score is relatively poor compared to the visual-oriented methods. Compared to MUNet-cWGAN and SRFlow, our method (Flow Enhancer) achieves better visual quality (LPIPS) while maintaining higher fidelity (PSNR and SSIM). Figure 2 shows the corresponding qualitative comparisons. MUNet provides SR images that are blurry compared to the ground truth images. MUNet-cWGAN improves the visual quality but tends to generate more artifacts than our Flow Enhancer. In addition, as shown in the first row (p1, tCh), Flow Enhancer recovers better contrast at the tumor than SRFlow and MUNet-cWGAN.

Ablation Studies. To justify our design, we performed ablation studies on 4 design elements: MRI anatomical information (MRI prior), conditional base distribution (Cond. Base), guide loss $\mathcal{L}_{guide}$ and DC loss $\mathcal{L}_{DC}$. Table 2 indicates that removing the MRI prior (in this case, $c = \{H\}$) harms all three metrics. Removing $\mathcal{L}_{guide}$ degrades the performance by a large margin. Removing the Cond. Base does not harm PSNR/SSIM but gives slightly worse LPIPS. $\mathcal{L}_{DC}$ imposes an L1 loss on the low-resolution components in k-space, therefore adding $\mathcal{L}_{DC}$ gives better PSNR/SSIM but slightly sacrifices the LPIPS (visual quality).

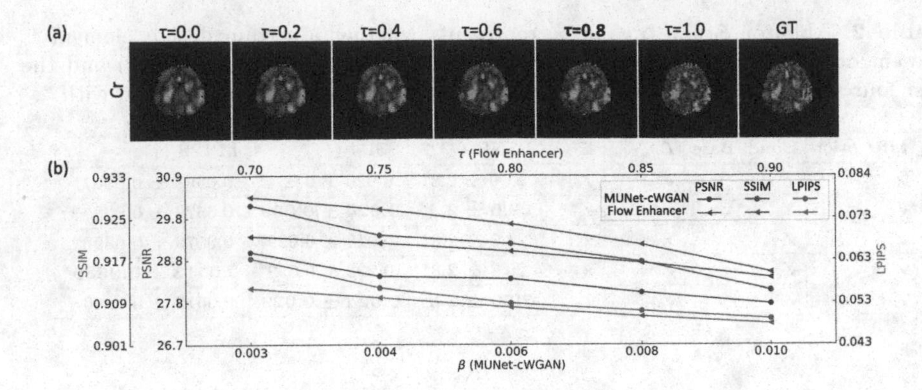

Fig. 3. Visual-fidelity tradeoff. (a) Sampling at different τ gives different levels of visual quality. $\tau = 0.8$ (bold) gives the closest visual quality as GT. (b) PSNR (↑), SSIM (↑) and LPIPS (↓) of Flow Enhancer and MUNet-cWGAN at various levels of visual quality.

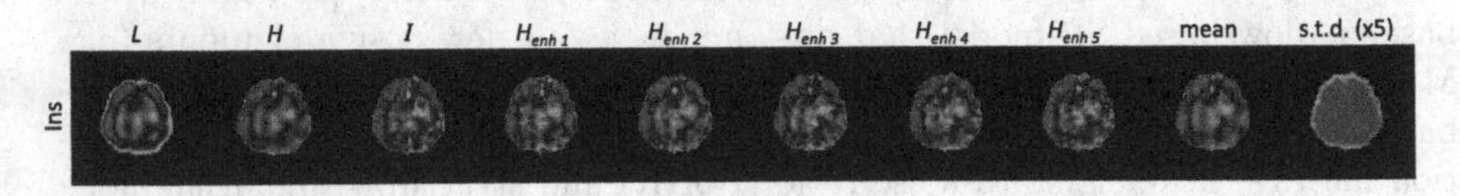

Fig. 4. Uncertainty estimation. From left to right: zero-filled low-resolution image L, blurry SR image H given by the MUNet, high-resolution ground truth I, 5 different samples $H_{enh1}, H_{enh2}, ..., H_{enh5}$ given by the enhancer network, mean and standard deviation (s.t.d.) calculated from 100 samples. The s.t.d. map is shown in $\times 5$ scale.

Controlling Visual-Fidelity Tradeoff via τ. Our enhancer network allows tuning the tradeoff between visual quality and fidelity within the same network via τ. A small value of τ gives blurry images with higher PSNR/SSIM. On the contrary, a large value of τ significantly improves the visual quality while sacrificing PSNR/SSIM. As shown in Fig. 3(a), $\tau = 0.8$ gives the closest visual quality as the ground truth, consistent with the recommendations in previous works [22,23]. We also compare our Flow Enhancer with MUNet-cWGAN at various levels of visual quality. For MUNet-cWGAN, the tradeoff is tuned via the weight of adversarial loss, i.e. β in $\mathcal{L}_{\text{MUNet-cWGAN}} = \mathcal{L}_{pixel} + \mathcal{L}_{structural} + \beta\mathcal{L}_{adversarial}$ [12]. Note that MUNet-cWGAN requires training a separated network for each β, whereas our Flow Enhancer only needs to be trained once to obtain different levels of visual quality. As shown in Fig. 3(b), Flow Enhancer achieves better visual quality (LPIPS) while maintaining higher image fidelity (PSNR/SSIM) compared to MUNet-cWGAN at all levels of visual quality.

Uncertainty Estimation. Different from the methods based on adversarial loss, flow-based models learn a target image manifold instead of a single solution. Sampling the latent variable $\mathbf{z} \sim \mathcal{N}(\boldsymbol{\mu}(c), \tau_0\boldsymbol{\sigma}(c))$ generates different samples from the learned image space, of which the standard deviation can be used for uncertainty estimation of the enhanced image H_{enh} [7]. Figure 4 shows an example of the standard deviation map calculated from 100 samples. The uncertainty is higher around the brain periphery, which means the network observes higher variances in these regions from the training dataset. This is probably due to the lower sensitivity and stronger spectra distortion near the skull. Note that the mean image is a pixelwise average of different samples in the learned image space, therefore it looks almost identical to the blurry SR image H.

4 Conclusion

We present a flow-based enhancer network to improve the visual quality of SR MRSI. Based on the SRFlow model, we incorporated MRI prior, learnable base distribution, guide loss and DC loss to boost the performance. Results show that our method outperforms the adversarial networks and the baseline flow-based methods. Our method also allows visual quality adjustment and uncertainty estimation. The method can be extended in the future to other modalities [11].

Acknowledgements. This work was supported by the NIH grant R01EB025840, R01CA206180 and R01NS035193. The data acquisition was supported by the Austrian Science Fund (FWF) grants KLI 646, P 30701 and P 34198.

References

1. Ardizzone, L., et al.: Analyzing inverse problems with invertible neural networks. arXiv preprint arXiv:1808.04730 (2018)
2. Blau, Y., Michaeli, T.: The perception-distortion tradeoff. In: Proceedings of the IEEE Conference on Computer Vision and Pattern Recognition, pp. 6228–6237 (2018)
3. Cengiz, S., Valdes-Hernandez, M.C., Ozturk-Isik, E.: Super resolution convolutional neural networks for increasing spatial resolution of ^{1}H magnetic resonance spectroscopic imaging. In: Valdés Hernández, M., González-Castro, V. (eds.) MIUA 2017. CCIS, vol. 723, pp. 641–650. Springer, Cham (2017). https://doi.org/10.1007/978-3-319-60964-5_56
4. Chen, S., Sun, S., Huang, X., Shen, D., Wang, Q., Liao, S.: Data-consistency in latent space and online update strategy to guide GAN for fast MRI reconstruction. In: Deeba, F., Johnson, P., Würfl, T., Ye, J.C. (eds.) MLMIR 2020. LNCS, vol. 12450, pp. 82–90. Springer, Cham (2020). https://doi.org/10.1007/978-3-030-61598-7_8
5. Coman, D., et al.: Extracellular pH mapping of liver cancer on a clinical 3T MRI scanner. Magn. Reson. Med. **83**(5), 1553–1564 (2020)
6. De Graaf, R.A.: In Vivo NMR Spectroscopy: Principles and Techniques. Wiley, Hoboken (2019)
7. Denker, A., Schmidt, M., Leuschner, J., Maass, P.: Conditional invertible neural networks for medical imaging. J. Imaging **7**(11), 243 (2021)
8. Dinh, L., Sohl-Dickstein, J., Bengio, S.: Density estimation using real NVP. arXiv preprint arXiv:1605.08803 (2016)

9. Dolz, J., Ben Ayed, I., Desrosiers, C.: Dense multi-path u-net for ischemic stroke lesion segmentation in multiple image modalities. In: Crimi, A., Bakas, S., Kuijf, H., Keyvan, F., Reyes, M., van Walsum, T. (eds.) BrainLes 2018. LNCS, vol. 11383, pp. 271–282. Springer, Cham (2018). https://doi.org/10.1007/978-3-030-11723-8_27
10. Dong, S., Chen, E.Z., Zhao, L., Chen, X., Liu, Y., Chen, T., Sun, S.: Invertible sharpening network for MRI reconstruction enhancement. arXiv preprint arXiv:2206.02838 (2022)
11. Dong, S., De Feyter, H.M., Thomas, M.A., de Graaf, R.A., Duncan, J.S.: A deep learning method for sensitivity enhancement in deuterium metabolic imaging (DMI). In: Proceedings of the 28th Annual Meeting of ISMRM. No. 0391 (2020)
12. Dong, S., et al.: High-resolution magnetic resonance spectroscopic imaging using a multi-encoder attention u-net with structural and adversarial loss. In: 2021 43rd Annual International Conference of the IEEE Engineering in Medicine & Biology Society (EMBC), pp. 2891–2895. IEEE (2021)
13. Dong, S., et al.: Multi-scale super-resolution magnetic resonance spectroscopic imaging with adjustable sharpness. arXiv preprint arXiv:2206.08984 (2022)
14. Hangel, G., et al.: High-resolution metabolic imaging of high-grade gliomas using 7T-CRT-FID-MRSI. NeuroImage: Clin. **28**, 102433 (2020)
15. He, K., Zhang, X., Ren, S., Sun, J.: Deep residual learning for image recognition. In: Proceedings of the IEEE Conference on Computer Vision and Pattern Recognition, pp. 770–778 (2016)
16. Iqbal, Z., Nguyen, D., Hangel, G., Motyka, S., Bogner, W., Jiang, S.: Super-resolution 1h magnetic resonance spectroscopic imaging utilizing deep learning. Front. Oncol. **9**, 1010 (2019)
17. Kelkar, V.A., Bhadra, S., Anastasio, M.A.: Compressible latent-space invertible networks for generative model-constrained image reconstruction. IEEE Trans. Comput. Imaging **7**, 209–223 (2021)
18. Kingma, D.P., Ba, J.: Adam: a method for stochastic optimization. arXiv preprint arXiv:1412.6980 (2014)
19. Kingma, D.P., Dhariwal, P.: Glow: generative flow with invertible 1x1 convolutions. Advances in Neural Information Processing Systems, vol. 31 (2018)
20. Li, W., et al.: Best-buddy GANs for highly detailed image super-resolution. arXiv preprint arXiv:2103.15295 (2021)
21. Li, X., et al.: Deep learning super-resolution MR spectroscopic imaging of brain metabolism and mutant IDH glioma. Neuro-Oncol. Adv. **4**(1) (2022)
22. Liang, J., Lugmayr, A., Zhang, K., Danelljan, M., Van Gool, L., Timofte, R.: Hierarchical conditional flow: a unified framework for image super-resolution and image rescaling. In: Proceedings of the IEEE/CVF International Conference on Computer Vision, pp. 4076–4085 (2021)
23. Lugmayr, A., Danelljan, M., Van Gool, L., Timofte, R.: SRFlow: learning the super-resolution space with normalizing flow. In: Vedaldi, A., Bischof, H., Brox, T., Frahm, J.-M. (eds.) ECCV 2020. LNCS, vol. 12350, pp. 715–732. Springer, Cham (2020). https://doi.org/10.1007/978-3-030-58558-7_42
24. Menon, S., Damian, A., Hu, S., Ravi, N., Rudin, C.: PULSE: self-supervised photo upsampling via latent space exploration of generative models. In: Proceedings of the IEEE/CVF Conference on Computer Vision and Pattern Recognition, pp. 2437–2445 (2020)
25. Padmanabha, G.A., Zabaras, N.: Solving inverse problems using conditional invertible neural networks. J. Comput. Phys. **433**, 110194 (2021)

26. Provencher, S.W.: Lcmodel & lcmgui user's manual. LCModel Version **6**(3) (2014)
27. Rezende, D., Mohamed, S.: Variational inference with normalizing flows. In: International Conference on Machine Learning, pp. 1530–1538. PMLR (2015)
28. Smith, S.M., et al.: Advances in functional and structural MR image analysis and implementation as FSL. Neuroimage **23**, S208–S219 (2004)
29. Wang, Z., Chen, J., Hoi, S.C.: Deep learning for image super-resolution: a survey. IEEE Trans. Pattern Anal. Mach. Intell. **43**(10), 3365–3387 (2020)
30. Wang, Z., Simoncelli, E.P., Bovik, A.C.: Multiscale structural similarity for image quality assessment. In: 2003 the Thrity-Seventh Asilomar Conference on Signals, Systems & Computers, vol. 2, pp. 1398–1402. IEEE (2003)
31. Winkler, C., Worrall, D., Hoogeboom, E., Welling, M.: Learning likelihoods with conditional normalizing flows. arXiv preprint arXiv:1912.00042 (2019)
32. Zhang, R., Isola, P., Efros, A.A., Shechtman, E., Wang, O.: The unreasonable effectiveness of deep features as a perceptual metric. In: Proceedings of the IEEE Conference on Computer Vision and Pattern Recognition, pp. 586–595 (2018)
33. Zhao, H., Gallo, O., Frosio, I., Kautz, J.: Loss functions for image restoration with neural networks. IEEE Trans. Comput. Imaging **3**(1), 47–57 (2016)

Cross Attention Transformers for Multi-modal Unsupervised Whole-Body PET Anomaly Detection

Ashay Patel(✉), Petru-Daniel Tudosiu, Walter Hugo Lopez Pinaya, Gary Cook, Vicky Goh, Sebastien Ourselin, and M. Jorge Cardoso

King's College London, London WC2R 2LS, UK
ashay.patel@kcl.ac.uk

Abstract. Cancers can have highly heterogeneous uptake patterns best visualised in positron emission tomography. These patterns are essential to detect, diagnose, stage and predict the evolution of cancer. Due to this heterogeneity, a general-purpose cancer detection model can be built using unsupervised learning anomaly detection models; these models learn a healthy representation of tissue and detect cancer by predicting deviations from healthy appearances. This task alone requires models capable of accurately learning long-range interactions between organs, imaging patterns, and other abstract features with high levels of expressivity. Such characteristics are suitably satisfied by transformers, and have been shown to generate state-of-the-art results in unsupervised anomaly detection by training on healthy data. This work expands upon such approaches by introducing multi-modal conditioning of the transformer via cross-attention, i.e. supplying anatomical reference information from paired CT images to aid the PET anomaly detection task. Using 83 whole-body PET/CT samples containing various cancer types, we show that our anomaly detection method is robust and capable of achieving accurate cancer localisation results even in cases where healthy training data is unavailable. Furthermore, the proposed model uncertainty, in conjunction with a kernel density estimation approach, is shown to provide a statistically robust alternative to residual-based anomaly maps. Overall, a superior performance is demonstrated against leading alternatives, drawing attention to the potential of these approaches.

Keywords: Transformers · Unsupervised anomaly detection · Cross-attention · Multi-modal · Vector quantized variational autoencoder · Whole-body · Kernel density estimation

1 Introduction

Positron Emission Tomography (PET) promises one of the highest detection rates for cancer amongst imaging modalities [14]). Through enabling the visu-

Supplementary Information The online version contains supplementary material available at https://doi.org/10.1007/978-3-031-18576-2_2.

A. Mukhopadhyay et al. (Eds.): DGM4MICCAI 2022, LNCS 13609, pp. 14–23, 2022.
https://doi.org/10.1007/978-3-031-18576-2_2

alization of metabolic activity, the efficacy of PET is brought down to the high metabolic rates of cancer cells [1]. By detecting changes on a cellular level, PET is ideal for detecting new and recurrent cancers [13]. In most clinical applications, however, PET is coupled with CT or MRI data to allow the incorporation of structural information with the results presented from PET imaging.

Cancer detection and segmentation present a wide range of clinically relevant tasks from staging, treatment planning, and surgical or therapy intervention planning. Although effective, PET imaging sensitivities can range as much as 35% depending on the cancer type and radiologist [17]. This can be of further issue in the case of metastatic cancer where dissemination can easily be overlooked in small, superficial lesions [20]. Considering these shortfalls, there is significant motivation for developing accurate automated detection methods.

Unsupervised methods have become an increasingly prominent field in recent years for automatic anomaly detection by eliminating the necessity of acquiring accurately labelled data [2,5]. These methods mainly rely on creating generative models trained on healthy data. Then during inference, anomalies are defined as deviations from the defined model of normality. This approach eliminates the requirement of labelled training data and generalises to unseen pathologies. However, its efficacy is often limited by the requirement of uncontaminated data with minimal anomalies present during training. The current state-of-the-art models for the unsupervised generative approach are held by the variational autoencoder (VAE) and its variants. In Baur et al. [2] VAE approach, the healthy data manifold is obtained by constraining the latent space to conform to that of a given distribution. The reconstruction error is then used to localise anomalies during inference. This approach, however, has limitations: from low fidelity reconstructions to the lack of resilience to reconstructing anomalous data.

To overcome some of these issues, an approach for unsupervised anomaly detection was presented utilising autoregressive models coupled with vector-quantised variational autoencoder (VQ-VAE) [15,18].

Transformers, who are currently state-of-the-art networks in the language modelling domain [22,25], use attention mechanisms to learn contextual dependencies regardless of location, allowing the model to learn long-distance relationships to capture the sequential nature of sequences. This general approach can be generalised to any sequential data, and many breakthroughs have seen the application of transformers in computer vision tasks [5,6,12,26]. Although having showcased state-of-the-art performance in unsupervised anomaly detection tasks for medical imaging data [21], these methods still rely on healthy data for model training. To the best of our knowledge, no prior research exists using unsupervised methods to accurately localise abnormalities while using training data containing anomalies. This task is important as it is often difficult or unethical to obtain healthy datasets of certain medical imaging modalities as some images are only acquired with prior suspicion of disease.

To address these problems, we propose a method for unsupervised anomaly detection and segmentation using multi-modal imaging via transformers with cross attention. This method is able to detect anomalies even when trained on anomalous data by leveraging the heterogeneity of metastatic cancer and anatomical information from CT. Furthermore utilising the generative aspect of the transformer model we propose and evaluate a kernel density estimation approach for generating a more robust alternative to residual based anomaly maps.

2 Background

The principal components behind the proposed whole-body anomaly detection model rely on using transformer models and auto-encoders to learn the probability density function of 3D whole-body PET scans. Although all training data contain anomalies, the spread of metastatic cancer and spatial distribution of anomalies across samples will result in such anomalies being unlikely, thus appearing at the likelihood tail-end of the learnt distribution. In order to use transformer models, images need to be expressed as a sequence of values, ideally categorical. As it is not computationally feasible to do this using voxel values, a compact quantized (discrete) latent space is used as input for the transformer via a VQ-GAN model [10,18] (a VQ-VAE with an adversarial component).

2.1 VG-GAN

The original VQ-VAE model [18] is an autoencoder that learns discrete latent representations of images. The model comprises of three principal modules: the encoder that maps a given sample $x \in \mathbb{R}^{H \times W \times D}$ onto a latent embedding space $\hat{z} \in \mathbb{R}^{h \times w \times d \times n_z}$ where n_z is the size of each latent vector. Each latent vector is quantized using an element-wise quantization of which each code $\hat{z}_{ijl} \in \mathbb{R}^{n_z}$ is mapped to its nearest vector $e_k, k \in 1, ...K$, where K is the vocabulary size of a codebook learnt jointly with model parameters. The final portion of the network is the decoder, which reconstructs the original observation from the quantized latent space. The discrete latent space representation is thus a sequence of indexes k for each code from the codebook. As autoencoders often have limited fidelity reconstructions [9], as proposed in [10], an adversarial component is added to the VQ-VAE network to form a VQ-GAN. Further formulations and architecture details can be found in Appendix B.

2.2 Transformer

Once a VQ-GAN model is trained on the entire training set containing anomalous data, the following stage is to learn the probability density function of the sequence of latent representations in an autoregressive manner. Transformer models rely on attention mechanisms to capture the relationship between inputs regardless of the distance or positioning relative to each other. Within each transformer layer, a self-attention mechanism is used to map intermediate representations with three vectors: query, key and value (see Appendix C for detailed formulation). This process, however, relies on the inner product between elements and as such, network sizing scales quadratically with sequence length. Given this limitation, achieving full attention with large medical data, even after the VQ-GAN encoding, comes at too high a computational cost. To circumvent this issue, many efficient transformer approximations have been proposed [7,24]. In this study, a Performer model is used; the Performer makes use of the FAVOR+ algorithm [7] which proposes a linear generalized attention that offers a scalable estimate of the attention mechanism. In using such a model, we can apply transformer-like models to much longer sequence lengths associated with whole-body data. In order to learn the probability density function of whole-body data, the discretised latent space z_q must take the form of a 1D sequence s using some arbitrary ordering. We then train the transformer model to maximise the training data's log-likelihood in an autoregressive manner. In doing so, the transformer learns the distribution of codebook indices for a given position i with respect to all previous inputs $p(s_i) = p(s_i \mid s_{<i})$.

3 Method

3.1 Anomaly Detection

To perform the baseline anomaly detection model on unseen data, first, we obtain the discrete latent representation of a test image using the VQ-GAN model. Next, the latent representation z_q is reshaped using a 3D raster scan into a 1D sequence s where the trained Performer model is used to obtain likelihoods for each latent variable. These likelihoods represent the probability of each token appearing at a given position in the sequence $p(s_i) = p(s_i \mid s_{<i})$, highlighting those of low probability of appearing in healthy data. Then tokens with likelihoods below an arbitrary threshold are selected to generate a binary resampling mask to indicate abnormal latent variables $p(s_i) < t$ (where t is a threshold determined empirically using a validation dataset; $t = 0.01$ was found to be optimal). Using the resampling mask, the latent variables are "healed" by resampling from the transformer and replacing them in the sequence. This approach replaces anomalous latent variables with those that are more likely to belong to a healthy distribution. Using the "healed" latent space, the VQ-GAN model reconstructs the original image x as a healed reconstruction x_r. Finally, a voxel-wise residual map can be calculated as $x - x_r$ with final segmentations calculated by thresholding the residual values. As areas of interest in PET occur as elevated uptake, residual maps are filtered to only highlight positive residuals.

3.2 CT Conditioning

There are often times when more information can be useful for inference. This can be in the imaging domain through multiple resolutions [4], or multiple modalities/spectrums [16]. It is for these tasks where cross-attention can prove beneficial. From a clinical point of view, whole-body PET scans are acquired in conjunction with MRI or CT data to provide an anatomical reference as structural information. Additionally, it can be observed that areas of high uptake are not always associated with anomalies. For example, areas of high metabolic activity like the heart, in addition to areas where radiotracer may collect like the kidney and bladder can show high uptake patterns. Sometimes these areas are not obvious from PET alone, and as such, the anatomical reference provided from CT data is beneficial. This leads to the main contribution of the work, namely anomaly detection incorporating CT data. This process works by generating a separate VQ-GAN model to reconstruct the PET-registered CT data. Then, both CT and PET data are encoded and ordered into a 1D sequence using the same rasterisation process, such that CT and PET latent tokens are spatially aligned. The transformer network is then adapted to include cross-attention layers [11] that feed in the embedded CT sequence after each self-attention layer. At each point in the PET sequence, the network has a full view of the CT data helping as a structural reference. In doing so, the problem of determining the codebook index at a given position i becomes $p(s_i) = p(s_i \mid s_{<i}, c)$ where c is the CT latent sequence (detailed formulation can be found in Appendix C). This approach, as visualised in Fig. 1 adds robustness to the anomaly detection framework by providing meaningful context in areas of greater variability in uptake that can be explained by the anatomical information within CT.

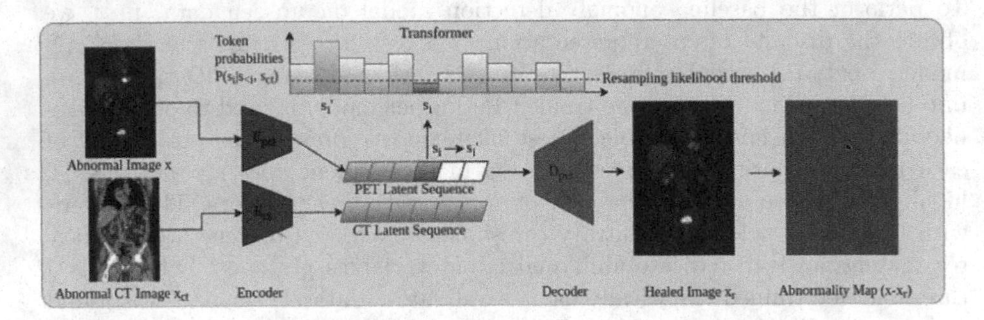

Fig. 1. Anomaly detection pipeline - PET image x is encoded along with CT image x_{ct}. Tokens from the encoded PET image are then sampled from the transformer by obtaining their likelihood with respect to prior tokens in the sequence and all CT tokens. Tokens below a given threshold are resampled from a multinomial distribution, derived from likelihood outputs from the transformer for all tokens at a given position in the sequence, giving a "healed" latent space which is decoded to give x_r.

3.3 Kernel Density Estimation

A drawback of the baseline anomaly detection method is that the residual image uses an arbitrary threshold to generate a segmentation map. The resulting segmentation can often be noisy due to discrepancies between the reconstructed image and the original, for example, between borders of high-intensity. Additionally, anomalies can occur at different intensities, meaning a blanket threshold is not appropriate. A possible solution to this is to implement Z-score anomaly maps as used in similar anomaly detection work [3]. For this work, this can be achieved by introducing variability within the model. However, certain uptake patterns can be related to base metabolic rate, in addition to procedure-related variations such as injected tracer amount and time since injection. As such, the optimality of the Z-score's Gaussian-error assumption should be questioned and likely relaxed. Empirical evidence obtained by exploring the data and by sampling from the transformer itself highlights that the error is indeed non-Gaussian even in healthy regions, for example the heart; bi-modal (even multi-modal) error distributions are observed. To remedy this, we propose to use a non-parametric approach using kernel density estimation (KDE) [19]. To do this, we introduce model uncertainty by using a dropout layer in the VQ-GAN decoder. Additionally, we obtain variability through replacing unlikely tokens with ones drawn from a multinomial distribution, derived from the likelihoods output from the transformer for each token at a given position in the sequence. By sampling multiple times, we generate multiple "healed" latent representations for a single image, which are then decoded multiple times with dropout to generate multiple "healed" reconstructions of a sample. At which point a KDE is fit at each voxel position to generate an estimate of the probability density function f. Letting $(x_1, \ldots, x_n)$ be the intensity for a voxel position across reconstructions, we can generate an estimation for the shape of the density function f for voxel x as:

$$\hat{f}_h(x) = \frac{1}{nh} \sum_{i=1}^{n} K\left(\frac{x - x_i}{h}\right) \tag{1}$$

where K is a gaussian kernel, and h is a smoothing bandwidth calculated as

$$h = \left(\frac{4\hat{\sigma}^5}{3n}\right)^{1/5} \tag{2}$$

with $\hat{\sigma}$ representing the standard deviation at a given voxel position across n reconstructions. We can then score voxels from that estimated density function at the intensity of the real image, at the voxel level, to generate a log-likelihood for that intensity, generating the anomaly map. To address areas of low variance across reconstructions, we implemented a minimum bandwidth of 0.05 (determined empirically using a validation dataset).

3.4 Clinically Consistent Segmentations for PET

For whole-body PET, the contours of an anomaly can be hard to define. The clinical standard in the UK defines boundaries of an anomaly as connecting voxels with intensities above 40% of the maximum intensity of a specific anomaly. To conform to this standard, we apply a final post-processing step of growing all initial segmentations to satisfy this criteria.

4 Results

The proposed models were trained on 60 images, with model and anomaly detection hyperparameter tuning carried out on 11 validation samples using the best

DICE scores. To assess our method's performance, we use 12 hold-out paired whole-body PET/CT images with varying cancers. We measure our models' performance using the best achievable DICE score, which serves as a theoretical upper-bound to the models segmentation performance. We obtained the scores using a greedy search approach for residual/density score thresholds. In addition, we calculate the area under the precision-recall curve (AUPRC), as a suitable measure for segmentation performance under class imbalance. We also compare our results to that of a VAE model proposed in [2]. Finally, we performed an ablation study of the proposed methods to demonstrate their added contribution along with paired t-tests to showcase the statistical significance of improvements.

Table 1. Anomaly detection results on whole-body PET data. The performance is measured with best achievable DICE-score ($\lceil DICE \rceil$) and AUPRC on the test set.

Method	$\lceil DICE \rceil$	AUPRC
VAE (Dense) [2]	0.359	0.282
VQ-GAN + Transformer (3D GAN variant of [21])	0.424	0.301
VQ-GAN + Transformer + CT conditioning (ours)	0.468	0.344
VQ-GAN + Transformer + CT conditioning + KDE (ours)	0.505	**0.501**
VQ-GAN + Transformer + CT conditioning + KDE + 40% Thresholding (ours)	**0.575**	0.458

Ablation Study: We observe a considerable improvement ($P = .001$) in anomaly detection performance by implementing CT conditioning in comparison to the 3D GAN variant approach of [21]. This result confirms our initial thoughts on the use case of anatomical context in the case of whole-body PET. Given the variability of healthy radiotracer uptake patterns, it is expected that beyond common areas like the bladder, further context is required to identify uptake as physiological or pathological. By incorporating model uncertainty to generate KDE maps, we see a further improvement in the overall DICE score, and even greater increase in AUPRC from 0.344 to 0.501 against the CT conditioned model ($P < .001$). This behaviour can be explained by the increased variability around heterogeneous areas of healthy uptake, attributing to a decrease in false positives. The main advantage of this approach, as visualised in Fig. 2 is the increase in precision. By discarding the assumption of Gaussian uptake distributions, the model can better differentiate patterns of physiological uptake from pathological whilst still being sensitive to subtle anomalies, as seen in sample C in Fig. 2.

Comparison to State-of-the-Art
From Table 1, we can see a statistically-significant improvement ($P = .001$) presented via the VQ-GAN + transformer approach using only PET data in relation to the VAE. This result is expected as demonstrated in prior research [21]. However, this divergence is also attributed to the presence of anomalies

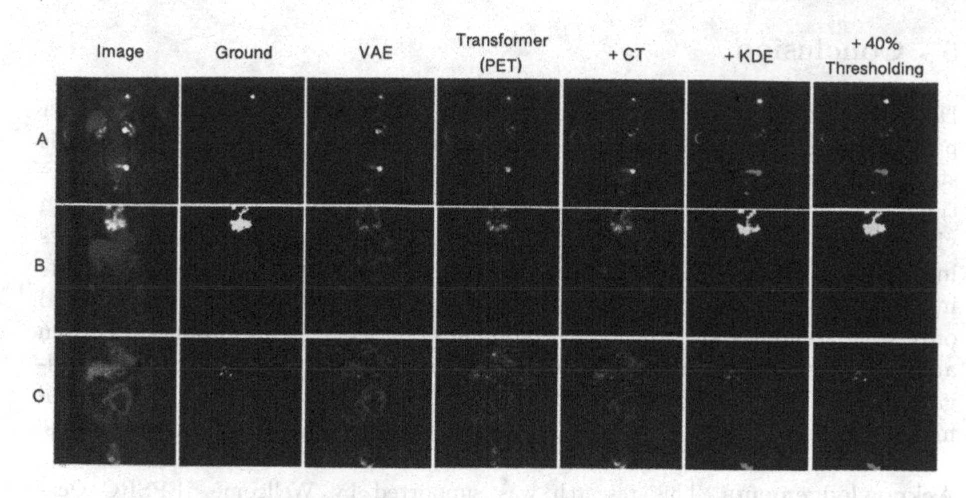

Fig. 2. Columns from left to right display (1st) the input image; (2nd) the gold standard truth segmentation; (3rd) the abnormality map as the residual for the VAE, (4th) Transformer, and (5th) CT conditioned methods; (6th) the abnormality map as a KDE, (7th) and after thresholding at 40% of each abnormal region maximum value. Results are provided for four randomly chosen subjects (A, B, C).

during training. It can be observed from sample B in Fig. 2, that the autoencoder method attempts to reconstruct large anomalies. Comparing the method proposed by [21] to our best model comprising of CT conditioning and KDE anomaly maps, our approach generates an improvement in DICE score from 0.424 to 0.505 ($P < .001$) with a considerable increase in AUPRC from 0.301 to 0.501 ($P < .001$). Finally, through clinically accurate segmentations by growing segmented regions, we see a large increase in the best possible DICE score, but a reduction in AUPRC brought about by the expansion of false-positive regions. From the results, there is clear evidence and motivation for the use of multi-modal conditioning for whole-body PET anomaly detection. In general from the qualitative results we can see detection results are high even from the PET only transformer approach however the incorporation of CT helps to improve precision through improved knowledge of the anatomical regions in the scan. Additionally the use of KDE based anomaly maps showcase a significant improvement on residual based maps. However, there are still areas for improvement beyond the current scope. We see varying cases of false positives across samples, showing ongoing difficulties differentiating physiological uptake from pathological. The reasons may be due to patient factors, i.e. general health, or more procedure-based factors, including radiotracer dosage and time since injection. Naturally, one solution would be to provide more training data; however, an alternative is to provide further conditioning related to the patient and procedure.

5 Conclusion

Detection and segmentation of anomalous regions, particularly for cancer patients, is essential for staging, treatment and intervention planning. In this study, we propose a novel pipeline for a transformer-based anomaly detection approach using multi-modal conditioning and kernel density estimation via model uncertainty. The model achieves statistically-significant improvements in Dice and AUPRC, representing a new state-of-the-art compared to competing methods. Additionally, we show the impact of this approach when faced only with training data containing anomalies, showing greater robustness than autoencoder only approaches. We hope that this work will inspire further investigation into anomaly detection with conditioned transformers using multi-modal medical imaging, and further exploration into the development of these methods.

Acknowledgements. This research was supported by Wellcome/ EPSRC Centre for Medical Engineering (WT203148/Z/16/Z), Wellcome Flagship Programme (WT213038/Z/18/Z), The London AI Centre for Value-based Heathcare and GE Healthcare. The models were trained on the NVIDIA Cambridge-1, UK's largest supercomputer, aimed at accelerating digital biology.

References

1. Almuhaideb, A., Papathanasiou, N., Bomanji, J.: 18F-FDG PET/CT imaging in oncology. Ann. Saudi Med. **31**, 3–13 (2011). https://doi.org/10.4103/0256-4947.75771
2. Baur, C., Denner, S., Wiestler, B., Albarqouni, S., Navab, N.: Autoencoders for unsupervised anomaly segmentation in brain MR images: a comparative study (2020)
3. Burgos, N., et al.: Anomaly detection for the individual analysis of brain pet images. J. Med. Imaging (Bellingham, Wash.) **8**, 024003 (2021). https://doi.org/10.1117/1.JMI.8.2.024003
4. Chen, C.F., Fan, Q., Panda, R.: CrossViT: cross-attention multi-scale vision transformer for image classification (2021)
5. Chen, M., Radford, A., Wu, J., Heewoo, J., Dhariwal, P.: Generative pretraining from pixels (2020)
6. Child, R., Gray, S., Radford, A., Sutskever, I.: Generating long sequences with sparse transformers (2019)
7. Choromanski, K., et al.: Rethinking attention with performers (2020)
8. Dhariwal, P., Jun, H., Payne, C., Kim, J.W., Radford, A., Sutskever, I.: Jukebox: a generative model for music (2020)
9. Dumoulin, V., et al.: Adversarially learned inference (2016)
10. Esser, P., Rombach, R., Ommer, B.: Taming transformers for high-resolution image synthesis (2020)
11. Gheini, M., Ren, X., May, J.: Cross-attention is all you need: Adapting pretrained transformers for machine translation (2021)
12. Jun, H., Child, R., Chen, M., Schulman, J.: Distribution augmentation for generative modeling (2020)

13. Kim, H.S., Lee, K.S., Ohno, Y., van Beek, E.J.R., Biederer, J.: PET/CT versus mri for diagnosis, staging, and follow-up of lung cancer. J. Magn. Reson. Imaging: JMRI **42**, 247–60 (2015). https://doi.org/10.1002/jmri.24776
14. Liu, B., Gao, S., Li, S.: A comprehensive comparison of CT, MRI, positron emission tomography or positron emission tomography/CT, and diffusion weighted imaging-MRI for detecting the lymph nodes metastases in patients with cervical cancer: A meta-analysis based on 67 studies. Gynecol. Obstet. Invest. **82**, 209–222 (2017). https://doi.org/10.1159/000456006
15. Marimont, S.N., Tarroni, G.: Anomaly detection through latent space restoration using vector-quantized variational autoencoders (2020)
16. Mohla, S., Pande, S., Banerjee, B., Chaudhuri, S.: FusAtNet: dual attention based spectrospatial multimodal fusion network for hyperspectral and lidar classification, pp. 416–425. IEEE (2020). https://doi.org/10.1109/CVPRW50498.2020.00054
17. Newman-Toker, D.E., et al.: Rate of diagnostic errors and serious misdiagnosis-related harms for major vascular events, infections, and cancers: toward a national incidence estimate using the "big three". Diagnosis (Berlin, Germany) **8**, 67–84 (2021). https://doi.org/10.1515/dx-2019-0104
18. van den Oord, A., Vinyals, O., Kavukcuoglu, K.: Neural discrete representation learning (2017)
19. Parzen, E.: On estimation of a probability density function and mode. Ann. Math. Stat. **33**, 1065–1076 (1962). https://doi.org/10.1214/aoms/1177704472
20. Perani, D., et al.: A survey of FDG- and amyloid-pet imaging in dementia and grade analysis. Biomed. Res. Int. **2014**, 785039 (2014). https://doi.org/10.1155/2014/785039
21. Pinaya, W.H.L., et al.: Unsupervised brain anomaly detection and segmentation with transformers (2021)
22. Radford, A., Narasimhan, K.: Improving language understanding by generative pre-training (2018)
23. Takaki, S., Nakashika, T., Wang, X., Yamagishi, J.: STFT spectral loss for training a neural speech waveform model (2018)
24. Tay, Y., et al.: Long range arena: a benchmark for efficient transformers (2020)
25. Vaswani, A., et al.: Attention is all you need (2017)
26. Yan, W., Zhang, Y., Abbeel, P., Srinivas, A.: VideoGPT: Video generation using VQ-VAE and transformers (2021)
27. Zhang, R., Isola, P., Efros, A.A., Shechtman, E., Wang, O.: The unreasonable effectiveness of deep features as a perceptual metric (2018)

Interpreting Latent Spaces of Generative Models for Medical Images Using Unsupervised Methods

Julian Schön[1,2(✉)], Raghavendra Selvan[1,3], and Jens Petersen[1,2]

[1] Department of Computer Science, University of Copenhagen, Copenhagen, Denmark
julian.e.s@di.ku.dk
[2] Department of Oncology, Rigshospitalet, Copenhagen, Denmark
[3] Department of Neuroscience, University of Copenhagen, Copenhagen, Denmark

Abstract. Generative models such as Generative Adversarial Networks (GANs) and Variational Autoencoders (VAEs) play an increasingly important role in medical image analysis. The latent spaces of these models often show semantically meaningful directions corresponding to human-interpretable image transformations. However, until now, their exploration for medical images has been limited due to the requirement of supervised data. Several methods for unsupervised discovery of interpretable directions in GAN latent spaces have shown interesting results on natural images. This work explores the potential of applying these techniques on medical images by training a GAN and a VAE on thoracic CT scans and using an unsupervised method to discover interpretable directions in the resulting latent space. We find several directions corresponding to non-trivial image transformations, such as rotation or breast size. Furthermore, the directions show that the generative models capture 3D structure despite being presented only with 2D data. The results show that unsupervised methods to discover interpretable directions in GANs generalize to VAEs and can be applied to medical images. This opens a wide array of future work using these methods in medical image analysis. The code and animations of the discovered directions are available online at https://github.com/julschoen/Latent-Space-Exploration-CT.

Keywords: Generative models · Unsupervised learning · Interpretability · CT

1 Introduction

The combination of deep learning and medical images has emerged as a promising tool for diagnostics and treatment. One of the main limitations is the often

Supplementary Information The online version contains supplementary material available at https://doi.org/10.1007/978-3-031-18576-2_3.

A. Mukhopadhyay et al. (Eds.): DGM4MICCAI 2022, LNCS 13609, pp. 24–33, 2022.
https://doi.org/10.1007/978-3-031-18576-2_3

small dataset sizes available for deep learning. Generative models can be used to mitigate this by synthesizing and augmenting medical images [12].

Generative Adversarial Networks (GANs) [6] have emerged as the prominent generative model for image synthesis. Consequently, research focusing on the interpretability of GANs has unfolded. At their inception, Radford et al. [20] showed meaningful vector arithmetic in the latent space of Deep Convolutional Generative Adversarial Networks (DCGANs). For several years, the methods used for discovering interpretable directions in latent spaces have been supervised [4,11,19] or based on simple vector arithmetic [20]. Especially in medical image analysis, supervision is expensive as it typically involves radiologists or other experts' time. Recently, several unsupervised methods for discovering interpretable directions in GAN latent spaces were proposed [7,23,25]. Due to being unsupervised, they seem more promising for the medical domain. However, it is still unclear if they work with the often more homogeneous images and the smaller dataset sizes encountered in this field.

Next to GANs, the interpretability of Variational Autoencoders (VAEs) [15] has also been studied extensively. However, the investigation has mainly focused on obtaining disentangled latent space representations [10,13]. While this shows promising results, it might not be possible without introducing inductive biases [17]. Applying the approaches for the unsupervised discovery of interpretable directions in latent spaces developed for GANs to VAEs might yield an alternative route for the investigation of interpretability in VAEs. Thus, if the same methods that have shown promising results on GANs are effective on VAEs, then VAEs can be trained without restrictions on the latent space, therefore not incorporating inductive biases while still having the benefit of interpretability and explicit data approximation.

Contributions: We employ a technique for the unsupervised discovery of interpretable directions in the latent spaces of DCGANs and VAEs trained on Computed Tomography (CT) scans. We show that these methods used to interpret the latent spaces of GANs generalize to VAEs. Further, our results provide insights into the applicability of these methods for medical image analysis. We evaluate the directions obtained and show that non-trivial and semantically meaningful directions are encoded in the latent space of the generative models under consideration. These directions include both transformations specific to our dataset choice and ones that likely generalize to other data. In particular, this allows for future work considering semantic editing of medical images in latent spaces of generative models.

2 Background

2.1 Generative Latent Models

As the backbone of this work we use generative latent models. We employ two of the most popular model types in GANs [6] for implicit and VAEs [15] for explicit approximation of the data distribution [5].

Given the discriminator D, the generator G, the latent distribution p_z, the data distribution p_{data}, and binary cross-entropy as the loss the GAN optimization is given by:

$$\min_G \max_D V(D,G) = \mathbb{E}_{x \sim p_{data}}[\log D(x)] + \mathbb{E}_{z \sim p_z}[\log(1 - D(G(z)))]. \quad (1)$$

We optimize the VAE using the Evidence Lower Bound (ELBO) with additional scaling factor β [10] given by:

$$\mathcal{L}_{VAE} = -\mathbb{E}_{q_\theta}[\log p_\phi(x|z)] + \beta D_{KL}[q_\theta(z|x)||p(z)] \quad (2)$$

where the first term is referred to as the reconstruction loss, with p_ϕ giving the likelihood parameterised by ϕ, and the second term as the regularization loss given by the Kullback-Leibler Divergence (KLD), with q_θ giving the approximate posterior parameterised by θ and $p(z)$ is the prior given by $p(z) \sim \mathcal{N}(0, I)$.

2.2 Discovery of Interpretable Directions in Latent Spaces

Several unsupervised methods to find interpretable directions in GAN latent spaces have been proposed [7,23,25]. In Härkönen et al.; Shen et al. [7,23] the directions are orthogonal. This constraint is relaxed in Voynov and Babenko [25]. As interpretable directions do not have to be orthogonal, we employ the method suggested by Voynov and Babenko [25]. The proposed method can be applied to any pretrained latent generative model G. The objective is to learn distinct directions in the latent space of G by learning a matrix A containing directions and a reconstructor R to distinguish between them. Since A and R are learned jointly, the directions of A are likely to be interpretable, semantically meaningful, and affect all images equally. Otherwise, distinguishing between the directions would be hard, and consequently, the accuracy of R would suffer.

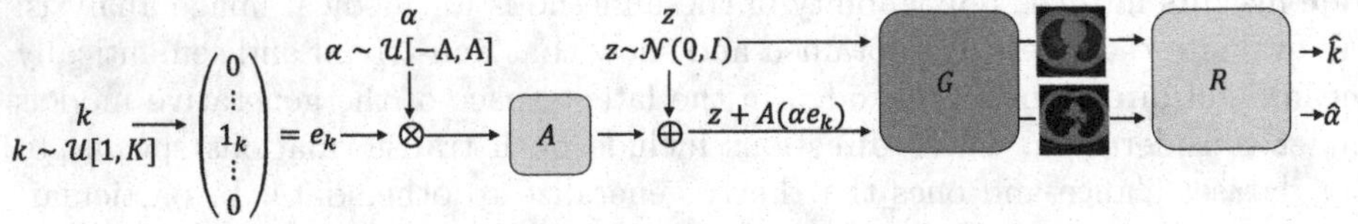

Fig. 1. Schematic overview of the learning protocol suggested by Voynov and Babenko. The upper path corresponds to the original latent code $z \sim \mathcal{N}(0, I)$ and the lower path corresponds to the shifted code $z + A(\alpha e_k)$ (Adapted from [25]).

Formally, the method learns a matrix $A \in \mathbb{R}^{d \times K}$, where d is the dimensionality of the latent space of G, and K is the number of directions that will be discovered. Thus, the columns of A correspond to discovered directions and are optimized during the training process to be easily distinguishable. Further, let $z \sim \mathcal{N}(0, I)$ be a latent code, e_k an axis-aligned unit vector with $k \in [1, ..., K]$

and α a scalar. Then, we can define the image pair $(G(z), G(z + A(\alpha e_k)))$ where $G(z)$ is the original image generated by latent code z and $G(z + A(\alpha e_k))$ is a shifted image from the original latent code z shifted along the kth discovered direction by amount α. Thus, α is a 'knob' controlling the magnitude of the shift. Given such an image pair, the method optimizes the reconstructor R presented with that pair to predict the shift direction k and amount α. Figure 1 illustrates the architecture. The optimization objective is given by:

$$\min_{A,R} \mathbb{E}_{z,k,\alpha}[L_{cl}(k, \hat{k}) + \gamma L_s(\alpha, \hat{\alpha})] \tag{3}$$

where k and α are the direction and amount respectively, and $\hat{k}$ and $\hat{\alpha}$ are the predictions. The classification term L_{cl} is given by cross-entropy. Further, we can use the classification term to get the Reconstructor Classification Accuracy (RCA), i.e., the accuracy of predicting the direction. Finally, the shift term L_s is given by the mean absolute error, and the regularization factor γ.

3 Material and Methods

3.1 Data

We use Lung Image Database Consortium image collection (LIDC-IDRI) [2] provided by The Cancer Imaging Archive (TCIA). It consists of clinical thoracic CT scans of 1010 patients collected from diagnostic and lung cancer screenings and is assembled by seven academic centers and eight medical imaging companies. We consider each axial slice as an individual image. Thus, our dataset consists of 246,016 CT slices. We resize the images to 128×128 pixels to limit computational demands and limited the data to a range of $[-1000, 2000]Hu$ to reduce the amount of outlier values and normalized using min-max scaling.

3.2 Models and Training

Since this study focuses on the potential of unsupervised exploration of latent spaces for medical images, we use simple generative models. We use a DCGAN based on Radford et al. [20], improving training stability by introducing one-sided label smoothing [22], replacing the fixed targets 1 of the real labels with smoothed values randomly chosen from the interval [0.9, 1]. Additionally, we add 0-mean and 0.1 standard deviation Gaussian noise to the discriminator input [1], incrementally reducing the standard deviation and finally removing it at the midpoint of training. The encoder and decoder of the VAE are based on ResNet [8], and we use $\beta = 0.01$ to improve reconstruction quality. For both generative models, we use a latent space size of $d = 32$ as it showed the best trade-off between image quality and compactness of the latent space. We refer to the provided GitHub repository for implementation details. We train the GAN and the VAE for 50 epochs selecting the best weights out of the last 5 by considering the models Fréchet Inception Distance (FID) [9] on test data. We use binary cross-entropy as loss for the GAN and log mean squared error [28] as reconstruction loss for the VAE. We use Adam [14] with a learning rate of 0.0002 and 0.0001 to optimize the GAN and VAE, respectively. The best model weights yield a FID of 33.4 for the GAN and 93.9 for the VAE on the test data.

To find interpretable latent directions, we use two different reconstructor architectures, based on LeNet [16] and ResNet18. We experiment with A having unit length or orthonormal columns as suggested by Voynov and Babenko [25]. We set the number of directions K equal to the size of the latent space, i.e., $K = 32$, and experiment with increasing it to $K = 100$. We observe significantly faster convergence when using the ResNet reconstructor. Thus, when using $K = 32$, we train the model for 25,000 iterations using LeNet and 3,000 iterations using the ResNet reconstructor. When $K = 100$, we train the VAE for 75,000 and 4,000 iterations with the LeNet and ResNet reconstructors respectively. For the GAN we observe slower convergence. Thus, we train for 250,000 and 10,000 iterations with the LeNet and ResNet reconstructors, respectively. Since we cannot have $K > d$ for orthonormal directions, we only use A with columns of unit length for $K = 100$. We evaluate direction models using the RCA and the shift loss L_s from Eq. 3. Further, we follow the ablation provided by Voynov and Babenko [25] and use a regularization factor $\gamma = 0.25$. To evaluate the directions, preliminary labeling was done by the first author with eight animations, each showing different latent vectors per direction. Next, each direction and preliminary label was considered on eight static images. The evaluator does not have formal training in medical image interpretation, and it is possible that more experienced evaluators could have discovered more interesting directions.

4 Experiments and Results

We perform several experiments to investigate the unsupervised exploration of latent spaces of deep generative models. First, we train using orthonormal directions and directions of unit length. We also experiment with increasing the number of directions. Finally, we perform all experiments both with a DCGAN and a VAE as generative models. All results are obtained without supervision, except the labeling of the selected directions. The RCA and Ls of the different experiments are presented in Table 1. We observe that the VAE always outperforms

Table 1. Reconstructor Classification Accuracy (RCA) and L_s for all model configurations for ResNet and LeNet as reconstructor.

	Orthogonal		Unit length		100 directions	
	RCA	L_s	RCA	L_s	RCA	L_s
GAN ResNet	0.9236	0.2538	0.9383	0.1949	0.9522	0.1560
GAN LeNet	0.8559	0.3317	0.9062	0.2439	0.9305	0.1406
VAE ResNet	0.9939	0.1040	0.9947	0.1086	0.9861	0.1117
VAE LeNet	0.9800	0.1421	0.9895	0.1090	0.9791	0.0962

the GAN with respect to both RCA and L_s. Further, using directions of unit length achieves higher RCA than orthonormal directions and lower L_s in all but one case. We also observe higher RCA when using ResNet over LeNet as a reconstructor. In contrast, LeNet achieves a lower L_s when K is set to 100.

Voynov and Babenko [25] mention that a larger K does not harm interpretability but alleviates entanglement and may lead to more duplicate directions. We observe the same behavior with $K = 100$ as opposed to $K = 32$.

Our results show eight consistent directions: width, height, size, rotation, y-position, thickness, breast size, and z-Position. All model configurations find all eight directions with varying degrees of entanglement. In this work, we omit directions entangled to such a degree that there is no clear interpretation dominating the image transformation. Thus, all configurations find at least a subset of the directions above in a sufficiently disentangled manner. We present animations of all discovered directions in the provided GitHub repository. Figure 2 shows all eight directions for the VAE and GAN. The directions presented are obtained using LeNet as reconstructor and $K = 100$. Directions obtained using different model configurations are presented in the supplementary material. Our results show that enforcing orthonormal directions increases entanglement. Finally, we

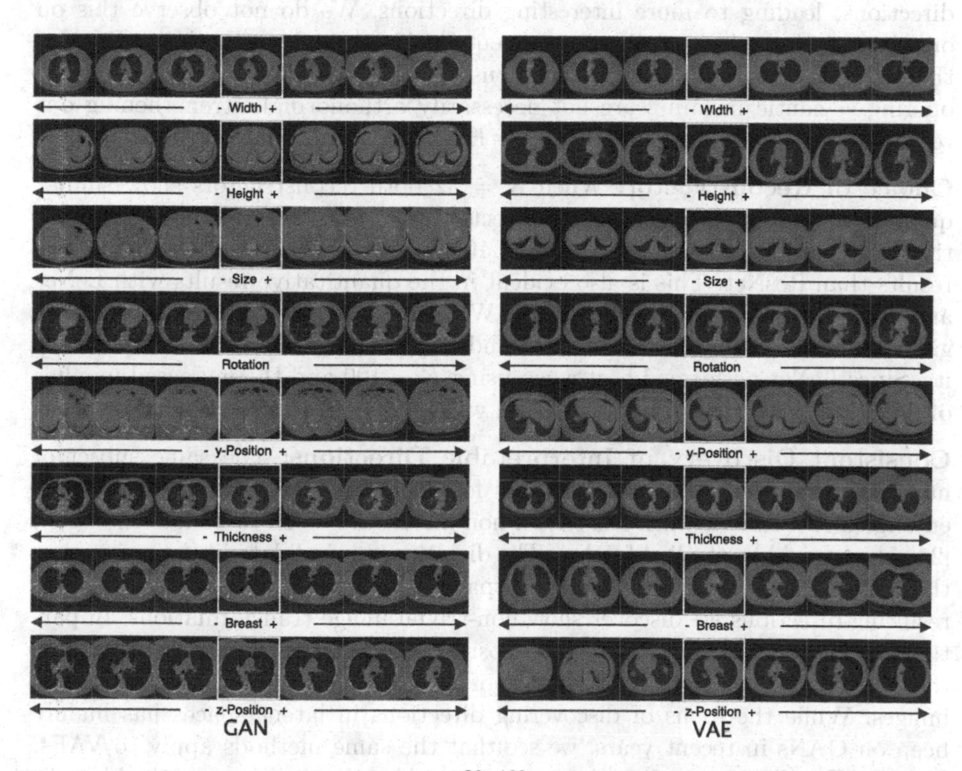

Fig. 2. Interpretable directions using $A^{32\times 100}$ with unit length columns, LeNet as reconstructor, and the GAN and VAE as generative models. The central images correspond to the original latent vector. The left/right images correspond to shifts.

observe that when using a LeNet reconstructor, more of the obtained directions are easily interpretable compared to using a ResNet reconstructor.

5 Discussion

In this work, we explored the latent spaces of deep generative models to discover semantically meaningful directions. We next elaborate on some of the findings of our experiments.

Influence of K: We observe less entanglement when increasing K. Thus, we hypothesize that lower K likely makes the reconstructor classification task easier, as there are less classes, lessening the need for disentanglement. If so, when increasing K, the increasing classification difficulty forces the model to disentangle the directions more.

Orthonormal Directions: While constraining the directions to be orthonormal still leads to the same subset of interpretable directions being discovered, their quality suffers. This aligns with the observations of Voynov and Babenko [25]. However, their results show that some datasets benefit from orthonormal directions, leading to more interesting directions. We do not observe this on our data, and the lack of disentanglement is also clear from the lower RCA of the methods using orthonormal directions. Thus, it seems likely that directions offering semantic meaning are not necessarily orthonormal, strengthening our reasoning for choosing this method over Härkönen et al.; Shen et al. [7,23].

Choice of Reconstructor: When $K = 32$ both reconstructors show similar qualitative results, more entangled directions, L_s is larger, and ResNet quantitatively outperforms LeNet. For $K = 100$, LeNet produces better qualitative results than ResNet. This is also evident in the quantitative results with LeNet and $K = 100$ achieving the lowest L_s. While ResNet has a higher RCA, RCA gives a measure of duplicate directions and only partially describes interpretability. Since LeNet performed best when using $K = 100$ and the increased number of directions benefited disentanglement, we prefer LeNet as reconstructor.

Consistent Discovery of Interpretable Directions: The same subset of human interpretable directions appears for all models with varying degrees of entanglement. Recent work has shown non-linear directions to be less entangled [24] which could be studied further. The directions are validated by showing that the same set is discovered in the latent space of both the generative models. The resulting directions we discover show non-trivial image transformations. In particular, the directions changing the z-Position of the latent vector demonstrates that the models learn the 3D structure of the data despite being trained on 2D images. While the focus of discovering directions in latent spaces has mainly been on GANs in recent years, we see that the same methods apply to VAEs. Since VAEs allow for explicit data approximation, they have a practical benefit over GANs when considering the usefulness of these methods.

Impact and Applications: Improving interpretability of GANs and VAEs is important and addressed in this work by finding and visualizing meaningful latent space directions and providing novel insights into the learned representations. The method is shown to generalize to VAEs, indicating that the latent spaces of VAEs and GANs can be interpreted in similar ways. However, shorter convergence times on the VAE when learning the directions indicate that VAEs latent spaces could be inherently easier to interpret. Unsupervised exploration further benefits the medical image domain due to the lack of well-supervised datasets, and more importantly, it could lead to surprising results outside of what we are explicitly supervising methods to find.

Our work can further be used for context-aware image augmentation and editing. Image augmentation using synthetic data improves downstream machine learning tasks on medical images [3] and can alleviate both the small dataset sizes and imbalance inherent to medical imaging [12,27]. Our results could be used to explore more diverse augmentations, e.g., adjusting for sex and weight imbalances. Additionally, our work might offer an alternate unsupervised approach to disease-aware image editing [21].

We see further applications needing more investigation, such as exploring the potential in consistency regularization and multi-modal datasets. For example, finding directions corresponding to adding or removing contrast in scans. Further, the approach we use has been shown to be effective in unsupervised saliency detection and segmentation on natural images [18,25,26].

Limitations: The main limitations we observe in our work are based on the methodology for unsupervised exploration. First, while the RCA and shift loss give some insights into convergence, the implications of overfitting need to be investigated. In particular, deciding how many training iterations to use is difficult as model performance can not be assessed on independent data. Further, the lack of evaluation metrics makes the choice of reconstructor difficult. We tried to mitigate this by using RCA and L_s for quantitative and human interpretation for qualitative analysis. Nevertheless, further investigation is needed to find good evaluation metrics. Second, the large amount of resulting directions makes evaluation difficult and time-consuming. This is particularly challenging in medical image analysis as evaluation may involve trained evaluators such as radiologists. Further automation or introducing a hierarchy of interpretability could be a focus of future work. Next to the methodological limitations, we see further potential for expanding our work to 3D generative models and more datasets in the future.

6 Conclusion

In this work, we have demonstrated for the first time that techniques for unsupervised discovery of interpretable directions in the latent space of generative models yield good results on medical images. While the interpretability of latent spaces is arguably an abstract concept depending on those interpreting, our results show that generative models learn non-trivial, semantically meaningful directions when trained on CT images of the thorax. We encounter directions with the same semantic meaning regardless of the generator or direction discovery model, indicating a general structure of the latent spaces. Further, our results show that the generative models' latent spaces capture the 3D structure of the CT scans despite only being trained on 2D slices. The work opens up the possibility of exploring these techniques for unsupervised medical image segmentation, interpolation, augmentation, and more.

Acknowledgements. The authors acknowledge the National Cancer Institute and the Foundation for the National Institutes of Health, and their critical role in the creation of the free publicly available LIDC/IDRI Database used in this study. The authors would like to thank Anna Kirchner and Arnau Morancho Tardà for help in preparation of the manuscript. Jens Petersen is partly funded by research grants from the Danish Cancer Society (grant no. R231-A13976) and Varian Medical Systems.

References

1. Arjovsky, M., Bottou, L.: Towards principled methods for training generative adversarial networks. In: 5th International Conference on Learning Representations (2017)
2. Armato III, S.G., et al.: The lung image database consortium (LIDC) and image database resource initiative (IDRI): a completed reference database of lung nodules on CT scans. Med. Phys. **38**(2), 915–931 (2011)
3. Chlap, P., Min, H., Vandenberg, N., Dowling, J., Holloway, L., Haworth, A.: A review of medical image data augmentation techniques for deep learning applications. J. Med. Imaging Radiat. Oncol. **65**(5), 545–563 (2021)
4. Goetschalckx, L., Andonian, A., Oliva, A., Isola, P.: GANalyze: toward visual definitions of cognitive image properties. In: IEEE/CVF International Conference on Computer Vision, pp. 5743–5752 (2019)
5. Goodfellow, I.J.: NIPS 2016 tutorial: generative adversarial networks. arXiv (2016)
6. Goodfellow, I.J., et al.: Generative adversarial nets. In: Advances in Neural Information Processing Systems 27: Annual Conference on Neural Information Processing Systems, pp. 2672–2680 (2014)
7. Härkönen, E., Hertzmann, A., Lehtinen, J., Paris, S.: GANspace: discovering interpretable GAN controls. In: Advances in Neural Information Processing Systems, vol. 33, pp. 9841–9850. Curran Associates, Inc. (2020)
8. He, K., Zhang, X., Ren, S., Sun, J.: Deep residual learning for image recognition. In: IEEE Conference on Computer Vision and Pattern Recognition, pp. 770–778. IEEE Computer Society (2016)
9. Heusel, M., Ramsauer, H., Unterthiner, T., Nessler, B., Hochreiter, S.: GANs trained by a two time-scale update rule converge to a local nash equilibrium. In: Advances in Neural Information Processing Systems, vol. 30. Curran Associates, Inc. (2017)
10. Higgins, I., et al.: beta-VAE: learning basic visual concepts with a constrained variational framework. In: 5th International Conference on Learning Representations (2017)
11. Jahanian, A., Chai, L., Isola, P.: On the "steerability" of generative adversarial networks. In: 8th International Conference on Learning Representations (2020)

12. Kazeminia, S., et al.: GANs for medical image analysis. Artif. Intell. Med. **109**, 101938 (2020)
13. Kim, H., Mnih, A.: Disentangling by factorising. In: Proceedings of the 35th International Conference on Machine Learning. Proceedings of Machine Learning Research, vol. 80, pp. 2649–2658 (2018)
14. Kingma, D.P., Ba, J.: Adam: a method for stochastic optimization. In: 3rd International Conference on Learning Representations (2015)
15. Kingma, D.P., Welling, M.: Auto-encoding variational bayes. In: 2nd International Conference on Learning Representations (2014)
16. Lecun, Y., Bottou, L., Bengio, Y., Haffner, P.: Gradient-based learning applied to document recognition. Proc. IEEE **86**(11), 2278–2324 (1998)
17. Locatello, F., et al.: Challenging common assumptions in the unsupervised learning of disentangled representations. In: Proceedings of the 36th International Conference on Machine Learning. Proceedings of Machine Learning Research, vol. 97, pp. 4114–4124 (2019)
18. Melas-Kyriazi, L., Rupprecht, C., Laina, I., Vedaldi, A.: Finding an unsupervised image segmenter in each of your deep generative models. arXiv (2021)
19. Plumerault, A., Le Borgne, H., Hudelot, C.: Controlling generative models with continuous factors of variations. In: International Conference on Machine Learning (2020)
20. Radford, A., Metz, L., Chintala, S.: Unsupervised representation learning with deep convolutional generative adversarial networks. In: 4th International Conference on Learning Representations (2016)
21. Saboo, A., Ramachandran, S.N., Dierkes, K., Keles, H.Y.: Towards disease-aware image editing of chest X-rays. arXiv (2021)
22. Salimans, T., et al.: Improved techniques for training GANs. In: Advances in Neural Information Processing Systems, vol. 29. Curran Associates, Inc. (2016)
23. Shen, Y., Zhou, B.: Closed-form factorization of latent semantics in GANs. In: IEEE/CVF Conference on Computer Vision and Pattern Recognition, pp. 1532–1540 (2021)
24. Tzelepis, C., Tzimiropoulos, G., Patras, I.: WarpedGANSpace: finding non-linear RBF paths in GAN latent space. In: IEEE/CVF International Conference on Computer Vision, pp. 6393–6402 (2021)
25. Voynov, A., Babenko, A.: Unsupervised discovery of interpretable directions in the GAN latent space. In: Proceedings of the 37th International Conference on Machine Learning. Proceedings of Machine Learning Research, vol. 119, pp. 9786–9796 (2020)
26. Voynov, A., Morozov, S., Babenko, A.: Object segmentation without labels with large-scale generative models. In: Proceedings of the 38th International Conference on Machine Learning. Proceedings of Machine Learning Research, vol. 139, pp. 10596–10606 (2021)
27. Yi, X., Walia, E., Babyn, P.S.: Generative adversarial network in medical imaging: a review. Med. Image Anal. **58**, 101552 (2019)
28. Yu, R.: A tutorial on VAEs: From bayes' rule to lossless compression. arXiv (2020)

What is Healthy? Generative Counterfactual Diffusion for Lesion Localization

Pedro Sanchez[1,2](✉), Antanas Kascenas[2,3], Xiao Liu[1,2], Alison Q. O'Neil[1,2], and Sotirios A. Tsaftaris[1,2,4]

[1] The University of Edinburgh, Edinburgh, Scotland, UK
pedro.sanchez@ed.ac.uk
[2] Canon Medical Research Europe, Edinburgh, Scotland, UK
[3] University of Glasgow, Glasgow, Scotland, UK
[4] The Alan Turing Institute, London, UK

Abstract. Reducing the requirement for densely annotated masks in medical image segmentation is important due to cost constraints. In this paper, we consider the problem of inferring pixel-level predictions of brain lesions by only using image-level labels for training. By leveraging recent advances in generative diffusion probabilistic models (DPM), we synthesize counterfactuals of "How would a patient appear if X pathology was not present?". The difference image between the observed patient state and the healthy counterfactual can be used for inferring the location of pathology. We generate counterfactuals that correspond to the minimal change of the input such that it is transformed to healthy domain. This requires training with healthy and unhealthy data in DPMs. We improve on previous counterfactual DPMs by manipulating the generation process with implicit guidance along with attention conditioning instead of using classifiers (Code is available at https://github.com/vios-s/Diff-SCM).

Keywords: Generative models · Diffusion probabilistic models · Counterfactuals

1 Introduction

Despite being crucial for training supervised machine learning models, pixel-level annotations of pathologies are costly. Creation of ground truth masks requires specialist radiologists. In this paper, we explore how to reduce the need for densely annotated ground truths in favour of a single image-level label: "Is the patient healthy or not?".

This problem has been tackled in the anomaly segmentation literature [7,11,16,20–23] by training deep generative models on healthy data only. They rely on the assumption that the model will learn "normal" (i.e. healthy) features whilst failing on out-of-distribution features [2]. In this case, a pixel-wise map

A. Mukhopadhyay et al. (Eds.): DGM4MICCAI 2022, LNCS 13609, pp. 34–44, 2022.
https://doi.org/10.1007/978-3-031-18576-2_4

can be generated by taking the residual between the input image and a predicted "healthy" image, to highlight the unhealthy areas. However, distinguishing normal (healthy) from abnormal (unhealthy) without being shown examples of abnormality is not trivial. In a brain lesion segmentation task [1], for instance, lesions deform adjacent areas of the brain. Arguably, these deformations should not be captured by the anomaly segmentation algorithm. In fact, it has been hypothesised that anomaly segmentation models [2] (trained only on healthy data) simply highlight zones of hyper-intensity in the image, since they may be surpassed in segmentation performance by simple thresholding techniques [8]. In line with this hypothesis, we show that, despite being more expressive than previous generative models,[1] diffusion probabilistic models (DPM) [3,4] trained on healthy data *only*, do *not* perform well in the segmentation task.

DPMs synthesise images by decomposing the generation process into a sequential application of denoising neural networks. We show empirically that efficient localization of brain lesions (abnormalities) with DPMs requires showing the model during training what an unhealthy brain is. Here, we demonstrate that the areas of interest (brain lesions) will be highlighted by performing the *minimal* intervention that can be applied to an image in order to change domains. This can be done using counterfactual generation [10,15]. In particular, a recent technique for generation of image counterfactuals [15] leverages a classifier (trained on image-level information) for manipulating an image between domains without paired domain data. The image is manipulated by encoding to latent space followed by conditional decoding. We create heatmaps by taking the difference between the observed image of a patient and its healthy counterfactual. However, using an extra classifier [15,18] can be cumbersome and hard to tune due to gradient dynamics during the iterative inference process of DPMs. In this paper, we improve on previous works by performing counterfactual diffusion without relying on downstream classifiers. We formulate a more efficient algorithm by using implicit guidance, attention-based conditioning and dynamic normalisation inspired by text-to-image DPMs [13,14].

Contributions. We explore brain lesion segmentation with generative DPMs without pixel-level supervision. We: (i) show that training DPMs on healthy data alone might not be sufficient for segmenting lesions, validating a previous hypothesis [8] that most anomaly segmentation algorithms only detect hyperintensities; (ii) perform counterfactual diffusion *without* relying on a downstream classifier, simplifying training and making the algorithm more robust to hyperparameter choices; and (iii) conduct extensive experiments, showing superior accuracy in brain lesion localization.

[1] Such as variational autoencoders (VAEs), normalizing flows (NFs) or generative adversarial networks (GANs).

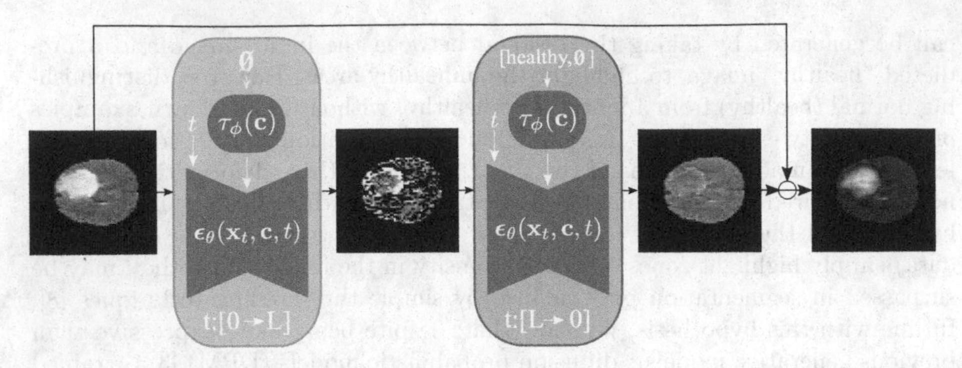

Fig. 1. Counterfactual diffusion overview. Encoding is done by iteratively applying diffusion models to obtain a latent space with an unconditional model ($\mathbf{c} = \emptyset$). **Decoding** is performed by reversing the diffusion process from the latent space to reconstruct an image with *healthy* condition. As detailed in Sect. 3.2, decoding uses *healthy* and $\emptyset$ for guidance. A heatmap of the lesion can obtained by taking the difference between the original and the reconstructed healthy.

2 Preliminaries on Diffusion Probabilistic Models (DPMs)

A diffusion process gradually adds noise to a data distribution over time. Diffusion probabilistic models (DPMs) [4] learn to reverse the noising process, going from noise towards the data distribution. DPMs can, therefore, be used as a generative model. The diffusion process gradually adds Gaussian noise, with a time-dependent variance α_t, to a data point $\mathbf{x}_0 \sim p_{\text{data}}(\mathbf{x})$ sampled from the data distribution. Thus, the noisy variable $\mathbf{x}_t$, with $t \in [0, T]$, is learned to correspond to versions of $\mathbf{x}_0$ perturbed by Gaussian noise following $p(\mathbf{x}_t \mid \mathbf{x}_0) = \mathcal{N}(\mathbf{x}_t; \sqrt{\alpha_t}\mathbf{x}_0, (1 - \alpha_t)\mathrm{I})$, where $\alpha_t := \prod_{j=0}^{t}(1 - \beta_j)$ and I is the identity matrix.

2.1 Training

With sufficient data and model capacity, the following training procedure ensures that the optimal solution to $\nabla_{\mathbf{x}} \log p_t(\mathbf{x})$ can be found by training $\boldsymbol{\epsilon}_\theta$ to approximate $\nabla_{\mathbf{x}} \log p_t(\mathbf{x}_t \mid \mathbf{x}_0)$ [6]. The diffusion model can be implemented with a conditional denoising U-Net $\boldsymbol{\epsilon}_\theta(\mathbf{x}_t, \mathbf{c}, t)$ which allows controlling the synthesis process through inputs $\mathbf{c}$. The training procedure is done learning a θ^* such that

$$\theta^* = \arg\min_{\theta} \mathbb{E}_{\mathbf{x}_0, t, \epsilon} \left[\|\boldsymbol{\epsilon}_\theta(\mathbf{x}_t, \mathbf{c}, t) - \epsilon\|_2^2 \right], \tag{1}$$

where $\mathbf{x}_t = \sqrt{\alpha_t}\mathbf{x}_0 + \sqrt{1 - \alpha_t}\epsilon$, with $\mathbf{x}_0 \sim p_{\text{data}}$ being a sample from the (training) data distribution, $t \sim \mathcal{U}(0, T)$ and $\epsilon \sim \mathcal{N}(0, \mathrm{I})$ is the noise.

2.2 Inference

Once the model ϵ_θ is learned using Eq. 1, generating samples consists in starting from $\mathbf{x}_T \sim \mathcal{N}(\mathbf{0}, \mathrm{I})$ and iteratively sampling from the reverse process with the diffusion model. Here, we will use the sampling procedure from Denoising Diffusion Implicit Models (DDIM, [17]) which formulates a deterministic mapping between latents to images following

$$\mathbf{x}_{t-1} = \sqrt{\alpha_{t-1}} \underbrace{\left(\frac{\mathbf{x}_t - \sqrt{1-\alpha_t} \cdot \epsilon_\theta(\mathbf{x}_t, \mathbf{c}, t)}{\sqrt{\alpha_t}} \right)}_{\hat{\mathbf{x}}_0} + \sqrt{1-\alpha_{t-1}}\, \epsilon_\theta(\mathbf{x}_t, \mathbf{c}, t). \quad (2)$$

The DDIM formulation has two main advantages: (i) it allows a near-invertible mapping between $\mathbf{x}_T$ and $\mathbf{x}_0$; and (ii) it allows efficient sampling with fewer iterations even when trained with the same diffusion discretization. This is obtained by choosing different under-sampling t in the $[0, T]$ interval.

3 Lesion Localization with Counterfactual Diffusion

We are interested in manipulating the input image from unhealthy[2] to healthy domain at inference time. At the same time, all other aspects of the input image should remain unchanged. Specifically, we are interested in identifying what is the main feature that should be modified. The main features should, for instance, correspond to lesions in a brain tumour dataset. This "minimal" manipulation is known in the causal literature as counterfactual generation [10,15]. Once $\epsilon_\theta(\mathbf{x}_t, \mathbf{c}, t)$ is trained on the imaging data with $\mathbf{c} \in$ [healthy, unhealthy] using Eq. 1, we can manipulate the input image between domains at inference time using the counterfactual generation inspired by [15,18].

3.1 Estimating Lesion Heatmap with Counterfactual Diffusion

We **encode** the input image into a (spatial) latent space by iteratively (L iterations) applying Eq. 2 in reverse order (simply changing $t-1$ to $t+1$) with an unconditional model ($\mathbf{c} = \emptyset$). Then, we generate a counterfactual by **decoding** the latent while applying an intervention to the conditioning $\mathbf{c}$ to be "healthy". Decoding is done by applying Eq. 2 with implicit guidance (using $\epsilon_\theta(\mathbf{x}_t, \mathbf{c}, t)$ as in Sect. 3.2) with attention conditioning (Sect. 3.3). The difference between the original image and counterfactual is then averaged along the channel dimension to obtain a heatmap which is used to recover the lesion (unhealthy features) segmentation. We apply dynamic normalization (Sect. 3.4) throughout the entire inference process. We illustrate this method on Fig. 1 and detail the algorithm in Algorithm 1.

[2] If the input is healthy, applying an intervention should not modify it.

3.2 Implicit Guidance

Using a classifier [3] to guide the diffusion process, which requires training an extra model over noisy images, has been successfully used to generate counterfactuals [15,18]. Here, we leverage *implicit guidance*[3] [5] in the context of counterfactual generation. In implicit guidance, a single diffusion model is trained on conditional and unconditional objectives via randomly dropping $\mathbf{c}$ during training (e.g. with 35% probability). The dropped conditioning is represented here with $\emptyset$ such that $\boldsymbol{\epsilon}_\theta(\mathbf{x}_t, \mathbf{c}, t)$ and $\boldsymbol{\epsilon}_\theta(\mathbf{x}_t, \emptyset, t)$ are conditional and unconditional $\boldsymbol{\epsilon}_\theta$-predictions. Sampling is performed by combining $\boldsymbol{\epsilon}_\theta$-predictions with a guidance *scale* (w), resulting in $\boldsymbol{\epsilon}_\theta(\mathbf{x}_t, \mathbf{c}, t) = w\boldsymbol{\epsilon}_\theta(\mathbf{x}_t, \mathbf{c}, t) + (1-w)\boldsymbol{\epsilon}_\theta(\mathbf{x}_t, \emptyset, t)$.

3.3 Conditioning with Attention Mechanisms

Generating counterfactuals requires conditioning the decoding during inference. As baseline, we utilise the adaptive group normalization (AdaGroup) which has already been successfully used in DPMs [3]. For counterfactual generation, modifying the normalization is not enough. We improve conditioning by augmenting the underlying U-Net backbone with a conditional attention mechanism inspired by previous work of text-to-image generation [9,12,13]. To pre-process $\mathbf{c}$, we use an encoder τ_ϕ that projects $\mathbf{c}$ to an intermediate representation $\tau_\phi(\mathbf{c}) \in \mathbb{R}^{d_\tau \times d_\tau}$, which is separately projected to the dimensionality of each attention layer throughout the model, and then concatenated to the attention context at each layer. In particular, we consider a U-Net with an attention layer implementing $\text{softmax}\left(\frac{QK_\mathbf{c}^T}{\sqrt{d}}\right) V_\mathbf{c}$. Similar to previous DPMs [3,4], the values for Q, K and V are derived from the previous convolutional layer. However, we concatenate $\tau_\phi(\mathbf{c})$ to K and V before the attention layer forming $K_\mathbf{c} = \text{concat}([K, \tau_\phi(\mathbf{c})])$ and $V_\mathbf{c} = \text{concat}([V, \tau_\phi(\mathbf{c})])$ [9].

3.4 Dynamic Normalization

During inference, the iterative process with guidance might change the input image statistics. This is specially cumbersome to counterfactual estimation because the latent space pixel values saturate (high absolute values). A saturated latent generates lower quality reconstructions and less manipulable images. Most DPM methods [3,9,13] clip the image to the $(-1, +1)$ range at each iteration. This *static* normalization ensures that the image can be appropriately processed by the neural network[4] but also results in a saturated latent. We avoid saturation by normalizing (dn), at each sampling step, the intermediate image to a certain percentile absolute pixel value. We use $th = \max(1, \text{percentile}(|\hat{x}_0|, s))$, where s is the desired percentage. Then, $\hat{x}_0$ is clipped to the range $(-th, th)$ and divided by th. This dynamic normalization pushes saturated pixels (those near -1 and $+1$) inwards, thereby actively preventing pixels from saturation at each step.

[3] Also known as *classifier-free* guidance in text-to-image generation DPMs [12,14].
[4] High absolute values of the neural network's input can result in unstable behaviour.

4 Experiments

4.1 Experimental Setup

Dataset. We evaluate the lesion localization performance on the surrogate task of brain tumor segmentation using data from the BraTS 2021 challenge [1]. This data comprises magnetic resonance (MR) imaging from four sequences T1, post-contrast T1-weighted (T1Gd), T2-weighted (T2), and T2 Fluid Attenuated Inversion Recovery (FLAIR) for each patient. The data has already been co-registered, skull-stripped and interpolated to the same resolution. We use dataset splits with 938, 62 and 251 patients for training, validation, and testing.

Training. The dataset has pixel-level annotations for the lesions. During training, we consider axial slices with at least one tumour voxel to be "unhealthy", and "healthy" otherwise. For the data input to the models, we concatenate all four modalities at the channel dimension for each patient. We normalize (rescale) the pixel values in each modality of each scan by dividing by the 99th percentile foreground voxel intensity. All slices are downsampled to a resolution of 64×64 for training, but all evaluation is done at 128×128 (1.62 mm/pixel) for fair comparison with baselines.

Benchmarks. We compare our model's performance against five generative methods, we use (i) a standard VAE [21,22];(ii) f-AnoGAN [16];(iii) VAE with iterative gradient-based restoration [20];(iv) denoising autoencoder (DAE) with coarse noise [7];(v) counterfactual diffusion model with classifier guidance [18][5]. Finally, we apply the simple thresholding approach from [8]. We use the hyperparameters from the original works for the deep learning methods but tune manually where necessary to improve training stability and performance. We detail in Table 1 if a benchmark method use only healthy data or the entire dataset during training.

Baseline. We use the denoising U-net $\boldsymbol{\epsilon}_\theta(\mathbf{x}_t, \mathbf{c}, t)$ from [3] as diffusion model and perform encoding-decoding inference as described in Sect. 3.1 as baseline entitled counterfactual DPM (CDPM). Following previous anomaly localization works [2,7], we also train a model $\text{CDPM}_{\text{healthy}}$ on healthy data only and inference is performed by encoding and decoding with an *unconditional* model $\boldsymbol{\epsilon}_\theta(\mathbf{x}_t, t)$. During the reconstruction process lesions should not be reconstructed because they are out-of-distribution.

Evaluation. We evaluate the lesion localization performance of the methods with two metrics (i) area under the precision-recall curve (AUPRC) at the pixel level computed for the whole test set; (ii) Dice score which measures the segmentation quality using the optimal threshold for binarization found by sweeping over possible values using the test ground truth. $\lceil \text{Dice} \rceil$ represents the upper bound for the Dice scores that would be obtainable in a more practical scenario.

[5] We train [18] at a different resolution than the original method for fair comparison. Therefore, we fine-tune their hyperparameters on a validation set for maximum performance as in Fig. 2.

4.2 Brain Lesion Localization

We now compare our method to previous benchmarks on brain lesion localization. As shown in Table 1, we surpass previous methods in a quantitative evaluation. In Fig. 3, we show the qualitative difference between the heatmaps created by our method when compared to other benchmarks. We also perform an ablation as shown in Table 2, studying the contribution of each individual components described in Sect. 3.

Table 1. Tumor detection performance as evaluated by test set wide pixel-level area under the precision-recall curve (AUPRC) and ideal Dice score (⌈Dice⌉). ± indicates standard deviation across 3 runs.

Method	AUPRC	⌈Dice⌉	Trained on
Thresholding [8]	68.4	66.7	Not
f-AnoGAN [16]	19.8 ± 0.6	31.6 ± 0.6	Healthy
VAE (reconstruction) [21,22]	29.9 ± 0.2	39.5 ± 0.2	Healthy
VAE (restoration) [20]	74.0 ± 0.7	68.5 ± 0.5	Healthy
DAE [7]	81.6 ± 0.5	75.8 ± 0.4	Healthy
$\text{CDPM}_{\text{healthy}}$	24.9 ± 0.4	33.1 ± 0.4	Healthy
CDPM + classifier [18]	81.5 ± 0.4	74.5 ± 0.4	Full
Ours	**82.8** ± 0.4	**76.2** ± 0.3	Full

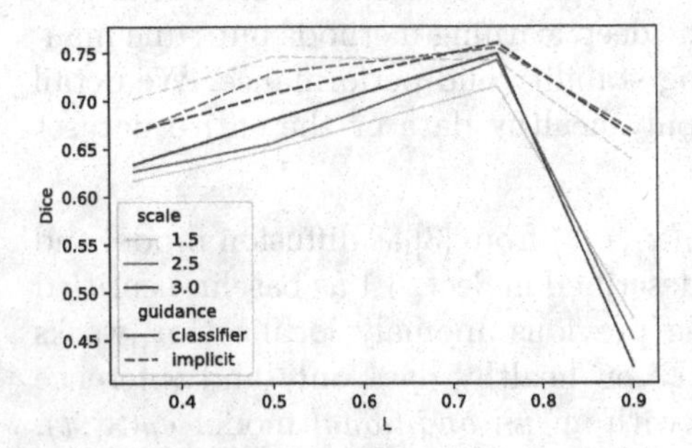

Fig. 2. Sweep through inference hyperparameters. We vary the guidance scale (both from classifier and implicit) and L defined in terms of percentage of training diffusion steps (1 corresponds to 100 DDIM steps).

Table 2. Contribution of each the components detailed throughout the paper to the localization results.

Improvements	Dice
$\text{CDPM}_{\text{healthy}}$	33.1
CDPM + Classifier [18]	74.5
CDPM	20.0
+ implicit guidance	52.0
+ attention conditioning	74.3
+ dynamic normalisation	**76.2**

4.3 How to Apply the Minimal Intervention?

The brain tumours are the main feature differentiating healthy from unhealthy images. Therefore, we explore how to ensure that lesions are highlighted and nothing else, resulting in higher Dice values. In our algorithm, the strength of the intervention can be controlled by varying the guidance *scale* (w) as well as the number of inference iterations L at inference time. We use a small (256 images) annotated validation set to find the best inference hyperparameters by measuring Dice while varying other variables as illustrated in Fig. 2. Our method ("implicit") is more robust to hyperparameter choice than a counterfactual diffusion using classifiers [18].

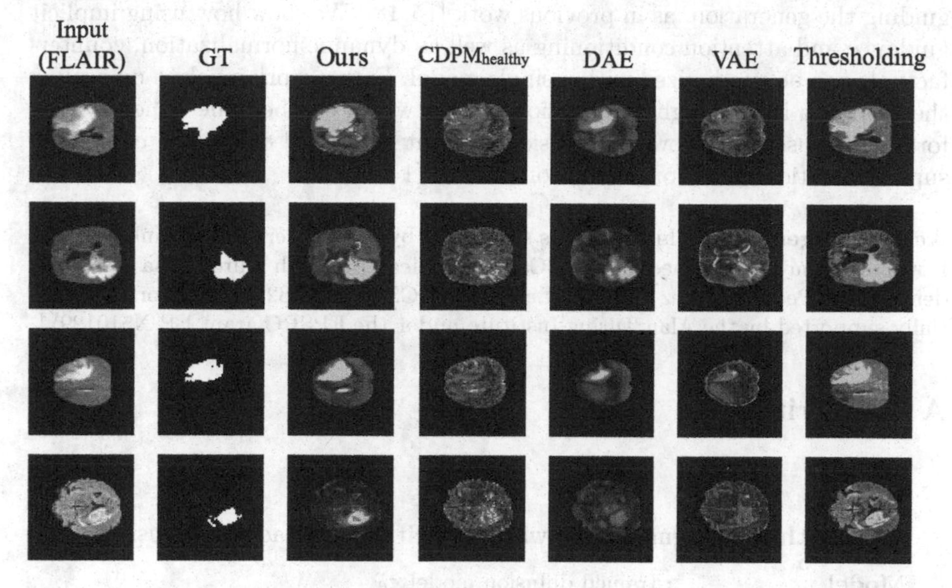

Fig. 3. Qualitative comparison of brain lesion segmentation methods. The columns indicate the input image (we are only illustrating the first channel but we use all four MRI sequences in our algorithm), the ground truth lesion masks and the heatmaps generated by each of the methods. Each row is a different slice.

5 Related Works on Generative Models for Lesion Localization

Variational autoencoders (VAEs) [21,22] constrain the latent bottleneck representation to follow a parameterized multivariate Gaussian distribution. [23] further add a context-encoding task and combine reconstruction error with density-based scoring to obtain the anomaly scores, while [20] use an iterative gradient descent restoration process at test time in restoration-VAE, replacing the reconstruction error with a restoration error to estimate anomaly scores. [16] train a generative adversarial network called f-AnoGAN which reuses the generator and discriminator to train an autoencoder with both reconstruction and adversarial losses for the anomaly detection task. Recently, [11] combine a vector quantized VAE (VQ-VAE) to encode an image with a DPM model over the latent variables in order to produce reconstructions with fewer reproduced anomalies. Other works explored pseudo-healthy pathology synthesis by disentangling representations in generative adversarial networks (GANs) [19].

6 Conclusions

We use conditional diffusion models for synthesizing healthy counterfactuals of a given input image, enabling lesion segmentation without access to pixel-level annotations. The difference between the observed image and the counterfactual produces a heatmap from which the segmentation masks can be obtained. Surprisingly, we show that this can be efficiently without a downstream classifier for guiding the generation, as in previous work [15,18]. We show how using implicit guidance and attention conditioning as well as dynamic normalization, counterfactuals can be synthesized with a single model. Future work involves up-scaling the model to handle higher resolution images which can be done either by performing diffusion in a lower dimensional latent space [13] or using a cascade of super-resolution conditional diffusion models [14].

Acknowledgements. This work was supported by the University of Edinburgh, the Royal Academy of Engineering and Canon Medical Research Europe via PhD studentships of Pedro Sanchez and Xiao Liu (grant RCSRF1819\825). This work was partially supported by the Alan Turing Institute under the EPSRC grant EP N510129\1.

A Algorithm

Algorithm 1: Segmentation with Implicit Counterfactual Diffusion

Model : trained diffusion model ϵ_θ
Hyper-parameters : guidance scale w, number of iterations L
Input : factual image $\mathbf{x}_0$ (M channels), condition $\mathbf{c}$
Output : heatmap

Recovering Unconditional Latent Space (Encoding)

for $t \leftarrow 0$ **to** L **do**

$$\hat{\mathbf{x}}_{t+1} = dn\left(\sqrt{\alpha_{t+1}}\left(\frac{\mathbf{x}_t - \sqrt{1-\alpha_t}\cdot\epsilon_\theta(\mathbf{x}_t, \emptyset, t)}{\sqrt{\alpha_t}}\right)\right) + \sqrt{\alpha_{t+1}}\epsilon_\theta(\mathbf{x}_t, \emptyset, t)$$

end

Counterfactual Generation (Decoding)

for $t \leftarrow L$ **to** 0 **do**

$$\epsilon = w\ \epsilon_\theta(\mathbf{x}_t, \mathbf{c}, t) + (1-w)\ \epsilon_\theta(\mathbf{x}_t, \emptyset, t)$$

$$\hat{\mathbf{x}}_{t-1} = dn\left(\sqrt{\alpha_{t-1}}\left(\frac{\mathbf{x}_t - \sqrt{1-\alpha_t}\cdot\epsilon}{\sqrt{\alpha_t}}\right)\right) + \sqrt{\alpha_{t-1}}\epsilon$$

end

$$\text{heatmap} = \frac{1}{M}\sum_m^M |\mathbf{x}_0^m - \hat{\mathbf{x}}_0^m|$$

References

1. Bakas, S., et al.: Identifying the best machine learning algorithms for brain tumor segmentation, progression assessment, and overall survival prediction in the BraTS challenge. arXiv preprint arXiv:1811.02629 (2018)
2. Baur, C., Denner, S., Wiestler, B., Navab, N., Albarqouni, S.: Autoencoders for unsupervised anomaly segmentation in brain MR images: a comparative study. Med. Image Anal. **69**, 101952 (2021)
3. Dhariwal, P., Nichol, A.Q.: Diffusion models beat GANs on image synthesis. In: Beygelzimer, A., Dauphin, Y., Liang, P., Vaughan, J.W. (eds.) Advances in Neural Information Processing Systems (2021)
4. Ho, J., Jain, A., Abbeel, P.: Denoising diffusion probabilistic models. In: Advances on Neural Information Processing Systems (2020)
5. Ho, J., Salimans, T.: Classifier-free diffusion guidance. In: NeurIPS 2021 Workshop on Deep Generative Models and Downstream Applications (2021)
6. Hyvärinen, A.: Estimation of non-normalized statistical models by score matching. J. Mach. Learn. Res. **6**, 695–709 (2005)
7. Kascenas, A., Pugeault, N., O'Neil, A.Q.: Denoising autoencoders for unsupervised anomaly detection in brain MRI. In: Medical Imaging with Deep Learning (2022)
8. Meissen, F., Kaissis, G., Rueckert, D.: Challenging current semi-supervised anomaly segmentation methods for brain MRI. In: Crimi, A., Bakas, S. (eds.) BrainLes 2021. LNCS, vol. 12962, pp. 450–462. Springer, Cham (2021). https://doi.org/10.1007/978-3-031-08999-2_5
9. Nichol, A., et al.: Glide: towards photorealistic image generation and editing with text-guided diffusion models. arXiv preprint arXiv:2112.10741 (2021)
10. Pawlowski, N., Castro, D.C., Glocker, B.: Deep structural causal models for tractable counterfactual inference. In: Advances in Neural Information Processing Systems (2020)
11. Pinaya, W.H., et al.: Fast unsupervised brain anomaly detection and segmentation with diffusion models. arXiv preprint arXiv:2206.03461 (2022)
12. Ramesh, A., Dhariwal, P., Nichol, A., Chu, C., Chen, M.: Hierarchical text-conditional image generation with clip latents. arXiv preprint arXiv:2204.06125 (2022)
13. Rombach, R., Blattmann, A., Lorenz, D., Esser, P., Ommer, B.: High-resolution image synthesis with latent diffusion models. In: Proceedings of the IEEE Conference on Computer Vision and Pattern Recognition (CVPR) (2022)
14. Saharia, C., et al.: Photorealistic text-to-image diffusion models with deep language understanding. arXiv preprint arXiv:2205.11487 (2022)
15. Sanchez, P., Tsaftaris, S.A.: Diffusion causal models for counterfactual estimation. In: First Conference on Causal Learning and Reasoning (2022)
16. Schlegl, T., Seeböck, P., Waldstein, S.M., Langs, G., Schmidt-Erfurth, U.: f-AnoGAN: fast unsupervised anomaly detection with generative adversarial networks. Med. Image Anal. **54**, 30–44 (2019)
17. Song, J., Meng, C., Ermon, S.: Denoising diffusion implicit models. In: Proceedings of International Conference on Learning Representations (2021)
18. Wolleb, J., Bieder, F., Sandkhler, R., Cattin, P.C.: Diffusion models for medical anomaly detection. arXiv preprint arXiv:2203.04306 (2022)
19. Xia, T., Chartsias, A., Tsaftaris, S.A.: Pseudo-healthy synthesis with pathology disentanglement and adversarial learning. Med. Image Anal. **64**, 101719 (2020)

20. You, S., Tezcan, K.C., Chen, X., Konukoglu, E.: Unsupervised lesion detection via image restoration with a normative prior. In: International Conference on Medical Imaging with Deep Learning, pp. 540–556. PMLR (2019)
21. Zhou, L., Deng, W., Wu, X.: Unsupervised anomaly localization using VAE and Beta-VAE. arXiv preprint arXiv:2005.10686 (2020)
22. Zimmerer, D., Isensee, F., Petersen, J., Kohl, S., Maier-Hein, K.: Unsupervised anomaly localization using variational auto-encoders. In: Shen, D., et al. (eds.) MICCAI 2019. LNCS, vol. 11767, pp. 289–297. Springer, Cham (2019). https://doi.org/10.1007/978-3-030-32251-9_32
23. Zimmerer, D., Kohl, S.A., Petersen, J., Isensee, F., Maier-Hein, K.H.: Context-encoding variational autoencoder for unsupervised anomaly detection. arXiv preprint arXiv:1812.05941 (2018)

Learning Generative Factors of EEG Data with Variational Auto-Encoders

Maksim Zhdanov[1,2](✉), Saskia Steinmann[3], and Nico Hoffmann[1]

[1] Helmholtz-Zentrum Dresden-Rossendorf, Dresden, Germany
maxxxzdn@gmail.com
[2] TU Dresden, Dresden, Germany
[3] Psychiatry Neuroimaging Branch, Department of Psychiatry and Psychotherapy, University Medical Center Hamburg-Eppendorf, Hamburg, Germany

Abstract. Electroencephalography produces high-dimensional, stochastic data from which it might be challenging to extract high-level knowledge about the phenomena of interest. We address this challenge by applying the framework of variational auto-encoders to 1) classify multiple pathologies and 2) recover the neurological mechanisms of those pathologies in a data-driven manner. Our framework learns generative factors of data related to pathologies. We provide an algorithm to decode those factors further and discover how different pathologies affect observed data. We illustrate the applicability of the proposed approach to identifying schizophrenia, either followed or not by auditory verbal hallucinations. We further demonstrate the ability of the framework to learn disease-related mechanisms consistent with current domain knowledge. We also compare the proposed framework with several benchmark approaches and indicate its classification performance and interpretability advantages.

Keywords: EEG · VAEs · Functional connectivity

1 Introduction

Analysis of neurological processes in the human brain is a challenging process addressed by neuroimaging. Here, one typically obtains high-dimensional stochastic data, which encourages the usage of machine learning algorithms. In recent years, deep learning discriminative models have been actively applied to neuroimaging issues (see [1] for a review). They yielded state-of-the-art results in classification problems on a variety of benchmark datasets [2–4]. One downside of deep learning-based classifiers is that they operate as black boxes [5] meaning that interpreting their predictions is often severely complicated.

S. Steinmann and N. Hoffmann—Equal contribution.

Supplementary Information The online version contains supplementary material available at https://doi.org/10.1007/978-3-031-18576-2_5.

A. Mukhopadhyay et al. (Eds.): DGM4MICCAI 2022, LNCS 13609, pp. 45–54, 2022.
https://doi.org/10.1007/978-3-031-18576-2_5

[1] suggested that hybrid generative-discriminative models might help resolve the issue. Such models can learn low-dimensional representations of data where each dimension corresponds to an independent *generative factor* (i.e. a disentangled representation, see [6] for a review). The discriminative part of the model then forces those factors to capture label information from data [7]. *Interpretability* is thus achieved via decoding the meaning of generative factors related to particular labels [8]. It is especially relevant in neuroimaging as one can observe how underlying pathologies govern the process of data generation.

The paper follows the intuition regarding hybrid generative-discriminative models for neuroimaging data, with particular application to EEG data. Our main contributions are as follows:

1. We demonstrate how one can apply characteristic capturing variational auto-encoders (CCVAEs) [7] to the interpretable classification of EEG data;
2. We compare the model to two generative models previously used for EEG data: conditional VAEs and VAEs with downstream classification;
3. We propose an algorithm for decoding generative factors learned by CCVAEs;
4. We demonstrated that learned generative mechanisms associated with pathologies are consistent with evidence from neurobiological studies.

2 Background

In this section, we introduce the relevant materials on variational auto-encoders, disentangled factorization and the role of supervision.

Variational Auto-Encoders. Variational auto-encoders (VAEs) [9] learn a model distribution $p_\theta(\mathbf{x}, \mathbf{z})$ that describes the ground-truth data generation process $p(\mathbf{x}, \mathbf{z})$ as first sampling random variables $\mathbf{z}$ from a prior distribution $p(\mathbf{z})$. Then, an observation $\mathbf{x}$ is inferred based on generative factors $p_\theta(\mathbf{x}|\mathbf{z})$ yielding

$$p_\theta(\mathbf{x}, \mathbf{z}) = p_\theta(\mathbf{x}|\mathbf{z})p(\mathbf{z}) \tag{1}$$

Here, the conditional distribution is parameterized with neural networks whose learned parameters are denoted with θ. Defining latent variables as jointly independent yields *disentangled factorization* [10] that separates the generative process into human-interpretable [8] generative mechanisms.

Supervised Learning. A label variable $\mathbf{y} \sim p(\mathbf{y})$ can be interpreted as the context that partially governs the generation of an observed variable $\mathbf{x}$. In VAEs, it is reflected by generative factors $p(\mathbf{x}|\mathbf{z}, \mathbf{y})$ of a model. It leads to the joint distribution factorized as follows:

$$p_{\theta_1,\theta_2}(\mathbf{x}, \mathbf{z}, \mathbf{y}) = p_{\theta_1}(\mathbf{x}|\mathbf{z}, \mathbf{y})p_{\theta_2}(\mathbf{z}|\mathbf{y})p(\mathbf{y}) \tag{2}$$

where θ_1, θ_2 are parameters of corresponding model distributions. The equality holds due to the chain rule. Incorporating label information into the model allows learning generative factors corresponding to those labels via $p_{\theta_2}(\mathbf{z}|\mathbf{y})$.

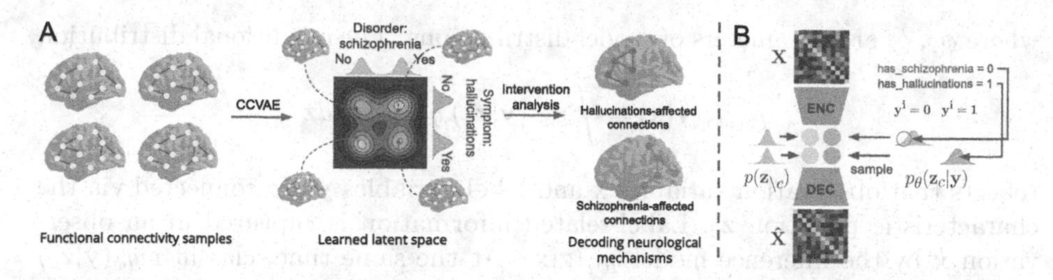

Fig. 1. Scheme of the proposed approach (A). We receive EEG data as input and learn a stochastic mapping to the latent space with CCVAEs (B) [7]. We further manipulate learned generative factors of data to gain insights regarding neurological mechanisms related to the attribute of interest, e.g. a symptom.

3 Methods

The proposed framework consists of 2 steps (see Fig. 1). First, EEG data is mapped stochastically to the latent space via CCVAEs. The latent space is constructed such that each label is related to a single independent generative factor. Second, we perform an intervention analysis to decode the meaning of label-related generative factors. This way, we get an intuition regarding mechanisms through which labels govern data generation. In our case, we are interested how different pathologies manifest themselves in functional connectivity matrices.

Characteristic Capturing VAEs. We aim at learning a model of a joint distribution over observed EEG data $\mathbf{x}$, labels (e.g. pathology indicators) $\mathbf{y}$ and latent variables $\mathbf{z}$ partially conditioned to $\mathbf{y}$. Let us assume that $\mathbf{x}$ and $\mathbf{y}$ are conditionally independent given $\mathbf{z}$. Then, the generative model (see Eq. 2) can be rewritten as follows:

$$p_{\theta_1,\theta_2}(\mathbf{x},\mathbf{y},\mathbf{z}) = p_{\theta_1}(\mathbf{x}|\mathbf{z})\, p_{\theta_2}(\mathbf{z}|\mathbf{y})\, p(\mathbf{y})$$

We further partition the latent space $\mathbf{z}$ such that one partition $\mathbf{z}_c$ encapsulates label associated characteristics, and the second partition $\mathbf{z}_{\setminus c}$ accounts for shared features of data (as in the vanilla VAEs):

$$p_{\theta_2}(\mathbf{z}|\mathbf{y}) = p_{\theta_2}(\mathbf{z}_c|\mathbf{y}) \cdot p(\mathbf{z}_{\setminus c})$$

The characteristic partition $\mathbf{z}_c$ is further partitioned so that each label can access only a single latent variable. It guarantees the disentanglement of label information in latent representations. The intractable distribution $p(\mathbf{z}|\mathbf{x},\mathbf{y})$ is conditioned to both observation and label variables. It is approximated with the following inference model:

$$q_{\phi_1,\phi_2}(\mathbf{z}|\mathbf{x},\mathbf{y}) = \frac{q_{\phi_1}(\mathbf{y}|\mathbf{z}_c)\, q_{\phi_2}(\mathbf{z}|\mathbf{x})}{q_{\phi_1,\phi_2}(\mathbf{y}|\mathbf{x})}$$

where ϕ_1, ϕ_2 are parameters of model distributions. The conditional distribution

$$q_{\phi_1,\phi_2}(\mathbf{y}|\mathbf{x}) = \int q_{\phi_1}(\mathbf{y}|\mathbf{z}_c)\, q_{\phi_2}(\mathbf{z}|\mathbf{x})\, d\mathbf{z}$$

reflects that observation variables $\mathbf{x}$ and label variables $\mathbf{y}$ are connected via the characteristic partition $\mathbf{z}_c$. Label-related information is captured in an observation $\mathbf{x}$ by the inference model $q_{\phi_2}(\mathbf{z}|\mathbf{x})$. At the same time, classifier $q_{\phi}(\mathbf{y}|\mathbf{z}_c)$ forces the label-related latent variables $\mathbf{z}_c$ to capture characteristics of those labels.

As for the vanilla VAEs, the model is optimized by maximizing the evidence lower bound [9]. In the case of CCVAEs, it is equivalent to maximizing the following objective (see Appendix B.1 of [7] for derivation):

$$\mathcal{L}(\mathbf{x}, \mathbf{y}) = \mathbb{E}_{q_{\phi_2}(\mathbf{z}|\mathbf{x})}\left[\frac{q_{\phi_1}(\mathbf{y}|\mathbf{z}_c)}{q_{\phi_1,\phi_2}(\mathbf{y}|\mathbf{x})} log \frac{p_{\theta_1}(\mathbf{x}|\mathbf{z})\, p_{\theta_2}(\mathbf{z}|\mathbf{y})}{q_{\phi_1}(\mathbf{y}|\mathbf{z}_c)\, q_{\phi_2}(\mathbf{z}|\mathbf{x})}\right] + log\, q_{\phi_1,\phi_2}(\mathbf{y}|\mathbf{x}) \quad (3)$$

The classification term $log\, q_{\phi_1,\phi_2}(\mathbf{y}|\mathbf{x})$ is essentially a learnable mapping from input data $\mathbf{x}$ to labels $\mathbf{y}$ that goes through the characteristic partition of the latent space $\mathbf{z}_c$. It applies pressure onto the partition to learn label-related characteristics from data and simultaneously performs data classification.

Intervention Analysis. The learned generative model forms the bridge between observations $\mathbf{x}$ and their labels $\mathbf{y}$ via latent variables $\mathbf{z}_c$. It allows one to analyze generative factors $p_{\theta_1,\theta_2}(\mathbf{x}|\mathbf{z}, \mathbf{y})$ of data related to those labels. One can explore the relation via intervention analysis. The algorithm for a single *binary* label of interest $\mathbf{y}^i$ is as follows. First, one fixes every dimension of the latent space $\mathbf{z}$ except the one $\mathbf{z}^i_c$ that corresponds to the label $\mathbf{y}^i$. Next, the value of $\mathbf{z}^i_c$ is sampled from $p_{\theta_2}(\mathbf{z}^i_c|\mathbf{y}^i)$ for each value of $\mathbf{y}^i \in \{0, 1\}$. As a result, one receives two latent representations $\mathbf{z}_0$, $\mathbf{z}_1$ that vary only in a single dimension $\mathbf{z}^i_c$. Those representations are then reconstructed to the observation space $\mathbf{x}_0 \sim p_{\theta_1}(\mathbf{x}|\mathbf{z} = \mathbf{z}_0)$, $\mathbf{x}_1 \sim p_{\theta_1}(\mathbf{x}|\mathbf{z} = \mathbf{z}_1)$. The procedure is repeated for N times. As a result, one gets multiple pairs of reconstructions $(\mathbf{x}_0, \mathbf{x}_1)$ that are different only to the varied generative factor $\mathbf{z}^i_c$. One further calculates the average difference $\frac{1}{N}\sum_{k=1}^{N}(\mathbf{x}^k_1 - \mathbf{x}^k_0)$ for each pair, and thus observes how the label $\mathbf{y}^i$ manifests itself in data.

4 Related Works

The fusion of generative and discriminative models with application to neuroimaging data is an active area of research. [11] demonstrate that using learned representations leads to more robust classification performance compared to feed-forward neural networks. [12] introduce VAEs into feature extraction from multichannel EEG data yielding better accuracy than traditional unsupervised approaches. [13] use stacked VAEs for semi-supervised learning on EEG data. However, the label information is usually encapsulated by multiple latent variables simultaneously. In this case, label characteristics are smeared across the latent space, thus complicating the analysis of label-related generative factors. It, in turn, limits both the interpretability and explainability of these models. One has to decode and interpret each latent variable and then infer the relation with label variables which is not a trivial task.

Two flavours of VAEs that are commonly applied to EEG data are conditional VAEs [13][1] and VAEs with downstream classification [12]. In both approaches, the latent space is not partitioned with respect to label variables. Hence, compared to CCVAEs, their general disadvantage is reduced interpretability of classification as it is difficult to build a bridge between labels and generative factors.

Conditional VAEs. Conditional VAEs have a graphical model similar to the one of CCVAEs. The only difference is that the latent space is not partitioned to labels, i.e. $\mathbf{z} = \mathbf{z_c}$. Learnable parameters are optimized via maximizing the objective Eq. (3). The framework allows conditional sampling, so one can use intervention analysis to decode the meaning of learned generative factors. Nevertheless, the interpretation is complicated as a single label variable is connected to each dimension of the latent space.

VAEs + Downstream Classification. The model approximates the joint distribution of observed data and latent variables that is factorized as Eq. (1). The relation between latent variables and labels is built via classifying a latent representation. The model is trained via optimizing the following objective [11]:

$$\mathcal{L}(\mathbf{x}, \mathbf{y}) = \mathbb{E}_{q_\phi(\mathbf{z}|\mathbf{x})}\Big[log\, p_\theta(x|z) - D_{KL}(q_\phi(z|x)||p(z)) - BCE(f_\xi(z), y)\Big] \quad (4)$$

where $f_\xi : Z \rightarrow Y$ is a learnable classifier with parameters ξ, BCE is binary cross-entropy function. Here, the information about label variables is incorporated into the latent space via pressure applied by a downstream classification task. The model can be seen as a feed-forward deep neural network with additional regularization imposed by the decoder part of VAEs.

5 Experimental Details

Experimental Study. The study comprised 29 patients suffering from schizophrenia and 52 healthy controls. 14 subjects out of those 29 indicated the emergence of auditory verbal hallucinations (AVH), i.e. hearing voices with no external stimuli presented. Every participant was right-handed. Six different

[1] Technically, [13] use stacked VAEs that have two connected latent spaces. One of the spaces is connected to label variables. However, the framework can be seen as an instance of conditional VAEs with a non-trivial structure of the latent space.

syllables were spoken to each participant (/ba/, /da/, /ka/, /ga/, /pa/, /ta/) for 500 ms simultaneously to each ear after 200 ms silence period. Meanwhile, the EEG recording was conducted with 64 electrodes where 4 EOG channels were used to monitor eye movements. For each subject, we repeated the procedure multiple times (number of trials for AVH: 68.23 ± 19.43; SZ: 68.76 ± 14.79 and HC: 71.19 ± 12.93). At the preprocessing step, the data was filtered from 20 120 Hz according to a protocol described in [14]. Therefore, only gamma-band frequencies are preserved. Afterwards, all channels were re-referenced to the common average. At last, muscle and visual artefacts were identified and removed. For our experiments, we utilized two parts of a recording: the resting one (first 200 ms with no syllable given) and the listening one (initial 200 ms when syllables were presented). The study of [14] contains detailed data acquisition and preprocessing information.

Experimental Data. For each EEG recording $[\zeta_1, \zeta_2, ..., \zeta_{61}]$, we assessed functional connectivity by calculating a correlation matrix:

$$\mathbf{x}_{ij} = \frac{cov(\zeta_i, \zeta_j)}{\sqrt{var(\zeta_i) \cdot var(\zeta_j)}}$$

As a result, functional connectivity matrices play the role of observed data $\mathbf{x}$. We introduce 3 binary labels such that $\mathbf{y} = [\textit{listening, schizophrenia, hallucinations}]$. To create a dataset for training models, we use the intra-patient paradigm, i.e. data from the same subject can appear simultaneously in training and test datasets. Thus, data of all the subjects are randomly sampled to form those datasets yielding 9000 training samples and 2000 test samples.

Implementation Details. One can find details regarding the parametrization of distributions in the supplementary material (Section S.1). We release the implementation at GitHub. For each framework, the parameters θ_i, ϕ_j (and ξ for VAEs) are trained via optimizing the corresponding objective. We use Adam optimizer with a learning rate of 10^{-3}. The training was performed in mini-batches of size 32 for 100 epochs. All models are trained on an NVIDIA Tesla V100 GPU from the Hemera HPC system of HZDR.

6 Results and Discussion

We found that high-dimensional latent spaces (dim > 32) hinder the reproducibility of generative factors learned by CCVAEs. For that reason, we keep the latent space of all models low-dimensional: $\mathbf{z} \in \mathbb{R}^5$ ($\mathbf{z}_c \in \mathbb{R}^3$, $\mathbf{z}_{\setminus c} \in \mathbb{R}^2$ for CCVAEs).

Results. As shown in Table 1, CCVAEs outperform baseline models in both classification performance and disentanglement (see supplementary material S.2 for details). The framework consistently classifies observed data based on its low-dimensional representation, yielding a low standard deviation of accuracy. Besides, it demonstrates a high level of disentanglement, meaning that each label variable is captured only by a single latent dimension. For CCVAEs, generative factors are disentangled in the latent space by design, leading to the highest score. It is not as high as expected due to the correlation between pathology labels.

Table 1. Comparison of CCVAEs to baseline models in terms of accuracy and disentanglement scores on the test dataset. For each framework, 10 experiments were conducted.

Framework	Accuracy	SAP score [16]	MIG score [15]
CCVAEs	**0.84 ± 0.01**	**0.34 ± 0.02**	**0.04 ± 0.01**
Conditional VAEs	0.69 ± 0.10	0.07 ± 0.07	0.01 ± 0.01
VAEs + classification	0.74 ± 0.03	0.06 ± 0.04	0.01 ± 0.01

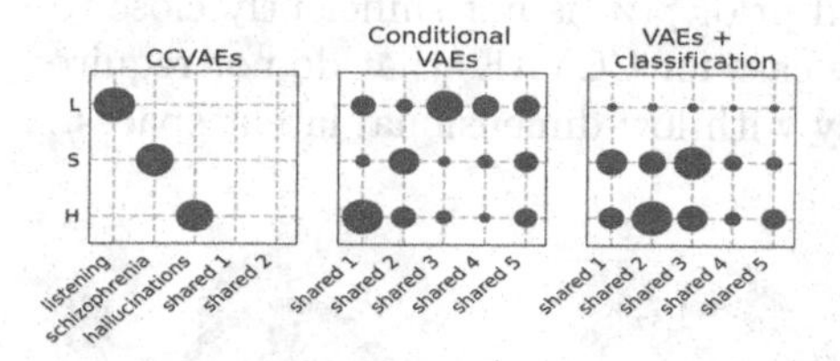

Fig. 2. Confusion matrices for different methods. The size of the circle indicates the value of the corresponding element. Rows correspond to label variables (L - *listening*, S - *schizophrenia*, H - *hallucinations*) while columns represent latent generative factors.

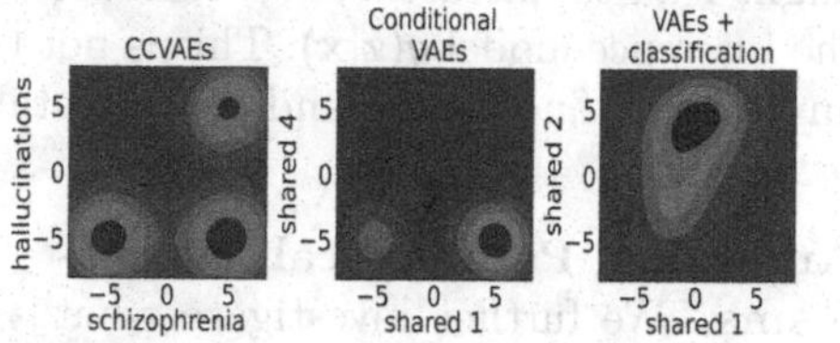

Fig. 3. Latent space generated by sampling from the inference model $q_\phi(\mathbf{z}|\mathbf{x})$ of different methods. For CCVAEs, the axes corresponding to pathology variables are shown. For baseline methods, 2 randomly selected dimensions are visualized.

Disentangled Latent Space. To demonstrate how hard-wired disentanglement affects the latent space learned by CCVAEs, we construct confusion matrices in the following way. We compute latent representation for each data point in the test dataset and intervene (i.e. randomly change its value) upon a single dimension. We further observe how log-probabilities assigned by a pre-trained classifier change due to the intervention for each label. We calculate the difference for each label-latent pair yielding a confusion matrix (see Fig. 2). The optimal result would be one non-zero element per row (i.e. label), which means that each label corresponds to only a single generative factor. This is the case of CCVAEs, where one can observe one-to-one dependence between label variables and corresponding generative factors. This leads to latent representations being robust to variations in data generative factors, as those are independent by design. At the same time, the characteristics of labels are entangled within latent spaces of baseline frameworks. Hence, it is difficult to disentangle the influence of a label from other generative factors, which severely hinders interpretability.

Posterior Distribution. We further compare latent spaces learned by each framework. To visualize the latent space, we sample $\mathbf{z} \sim q_\phi(\mathbf{z}|\mathbf{x})$ for multiple $\mathbf{x}$ from the test dataset for each model. The result is shown in Fig. 3. In the case of CCVAEs, the distribution has three modes corresponding to subject cohorts in data (healthy, schizophrenia, schizophrenia followed by AVH). The separation is caused by the influence of the conditional prior and the classifier, aiming to separate representations encoding different label combinations. In the case of baseline frameworks, there is no strict regularisation that preserves label information within a partition of the latent space. As a result, features of data encoded by their latents are shared between cohorts of subjects (thus one or two modes). We discovered that both baseline methods often fail to jointly learn a low-dimensional representation and classify labels when the pressure on the KL divergence term in the loss objective is high. The problem is partially solved by introducing a scaling factor β for the term [8]. However, reducing the pressure might lead to untrustable reconstruction if prior $p(\mathbf{z})$ is not sufficiently close to the inference model $q(\mathbf{z}|\mathbf{x})$. This is not the case for CCVAEs that do not require any manual fine-tuning and operate stably with low-dimensional latent spaces.

Analyzing Pathological Mechanisms. We further investigate what connections are affected when intervening upon a single label dimension via intervention analysis (Fig. 4, see supplementary material S.3 for computation details). The model associates the emergence of AVH with alterations in frontotemporal brain areas (the highest positive difference), which have been repeatedly observed in prior studies [17,18]. The salient connections are mainly located in the right hemisphere, which is supported by the fMRI study of [19]. The model also points toward reduced connectivity between hemispheres. It is coherent with the current hypothesis (see [14] for review) that connects the emergence of auditory verbal hallucinations with the interhemispheric miscommunication during auditory processing. Overall, the model can at least partially reconstruct the neurological mechanism of the symptom for functional connectivity. To explain the emergence of schizophrenia, the model focuses mainly on the left hemisphere. It is not surprising since the auditory function is left-lateralized for right-handed people [20,21]. It would be an interesting direction for further studies to apply CCVAEs to learn the mechanisms of particular symptoms of the composite disorder (e.g. hallucinations, delusions, etc.).

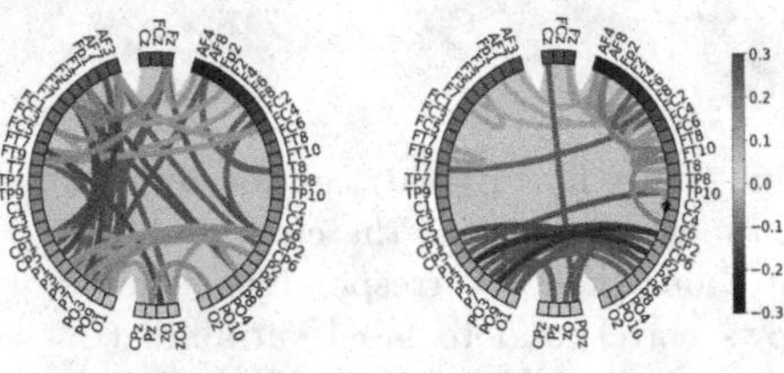

$\mathbf{y}^i$ = schizophrenia $\quad$ $\mathbf{y}^i$ = hallucinations

Fig. 4. Average difference in reconstructions of functional connectivity matrices when intervening on a single label: *schizophrenia* (Left) and *hallucinations* (Right). Connections that are stronger when a disorder is presented are shown in red; otherwise, blue. For clarity, we visualize only 40 connections with the highest absolute value.

7 Conclusion

We demonstrated how to apply the framework of characteristic capturing variational auto-encoders to EEG data analysis. The method encapsulates and disentangles the characteristics associated with different pathologies in the latent space. As generative factors are independent by design, one can decode their meaning and discover how those pathologies alter observed data. It leads to improved interpretability coupled with the high classification performance of neural networks. The framework is not limited to functional connectivity analysis or EEG data and can be easily adapted to different neuroimaging modalities.

References

1. Davatzikos, C.: Machine learning in neuroimaging: progress and challenges. Neuroimage **197**, 652–656 (2019)
2. Zhang, L., Wang, M., Liu, M., Zhang, D.: A survey on deep learning for neuroimaging-based brain disorder analysis. Front. Neurosci. **14**, 779 (2020)
3. Wahlang, I., et al.: Brain magnetic resonance imaging classification using deep learning architectures with gender and age. Sensors **22**, 1766 (2022)
4. Li, Y., et al.: A novel bi-hemispheric discrepancy model for EEG emotion recognition. IEEE Trans. Cogn. Dev. Syst. **13**, 354–367 (2021)
5. Quinn, T.P., Jacobs, S., Senadeera, M., Le, V., Coghlan, S.: The three ghosts of medical AI: can the black-box present deliver? Artif. Intell. Med. **124**, 102158 (2022)
6. Liu, X., Sanchez, P., Thermos, S., O'Neil, A.Q., Tsaftaris, S.A.: Learning disentangled representations in the imaging domain. Med. Image Anal. (2022)
7. Joy, T., Schmon, S.M., Torr, P.H., Siddharth, N., Rainforth, T.: Capturing label characteristics in VAEs. In: ICLR (2021)
8. Higgins, I., et al.: Beta-VAE: learning basic visual concepts with a constrained variational framework. In: ICLR (2017)
9. Kingma, D.P., Welling, M.: Auto-Encoding Variational Bayes. CoRR, abs/1312.6114 (2014)
10. Locatello, F., Bauer, S., Lucic, M., Gelly, S., Schölkopf, B., Bachem, O.: Challenging Common Assumptions in the Unsupervised Learning of Disentangled Representations. arXiv, abs/1811.12359 (2019)
11. Krishna, G., Tran, C., Carnahan, M., Tewfik, A.H.: Constrained Variational Autoencoder for improving EEG based Speech Recognition Systems. arXiv, abs/2006.02902 (2020)
12. Li, X., et al.: Latent factor decoding of multi-channel EEG for emotion recognition through autoencoder-like neural networks. Front. Neurosci. **14**, 87 (2020)
13. Chen, J., Yu, Z., Gu, Z.: Semi-supervised Deep Learning in Motor Imagery-Based Brain-Computer Interfaces with Stacked Variational Autoencoder (2020)
14. Steinmann, S., Leicht, G., Andreou, C., Polomac, N., Mulert, C.: Auditory verbal hallucinations related to altered long-range synchrony of gamma-band oscillations. Sci. Rep. **7**(1), 1–10 (2017)
15. Chen, T.Q., Li, X., Grosse, R.B., Duvenaud, D.K.: Isolating sources of disentanglement in variational autoencoders. In: NeurIPS (2018)
16. Kumar, A., Sattigeri, P., Balakrishnan, A.: Variational Inference of Disentangled Latent Concepts from Unlabeled Observations. arXiv, abs/1711.00848 (2018)

17. Jardri, R., Pouchet, A., Pins, D., Thomas, P.: Cortical activations during auditory verbal hallucinations in schizophrenia: a coordinate-based meta-analysis. Am. J. Psychiatry **168**(1), 73–81 (2011)
18. Lavigne, K.M., et al.: Left-dominant temporal-frontal hypercoupling in schizophrenia patients with hallucinations during speech perception. Schizophr. Bull. **41**(1), 259–67 (2015)
19. Hwang, M., et al.: Auditory hallucinations across the psychosis spectrum: evidence of dysconnectivity involving cerebellar and temporal lobe regions. NeuroImage Clin. **32**, 102893 (2021)
20. Papathanassiou, D., Etard, O., Mellet, E., Zago, L., Mazoyer, B., Tzourio-Mazoyer, N.: A common language network for comprehension and production: a contribution to the definition of language epicenters with PET. Neuroimage **11**, 347–357 (2000)
21. Flinker, A., et al.: Redefining the role of Broca's area in speech. Proc. Natl. Acad. Sci. **112**, 2871–2875 (2015)

An Image Feature Mapping Model for Continuous Longitudinal Data Completion and Generation of Synthetic Patient Trajectories

Clément Chadebec[1], Evi M. C. Huijben[2(✉)], Josien P. W. Pluim[2], Stéphanie Allassonnière[1], and Maureen A. J. M. van Eijnatten[2]

[1] Université Paris Cité, INRIA, Inserm, Sorbonne Université, Paris, France
clement.chadebec@inria.fr

[2] Department of Biomedical Engineering, Medical Image Analysis Group, Eindhoven University of Technology, Eindhoven, The Netherlands
e.m.c.huijben@tue.nl

Abstract. Longitudinal medical image data are becoming increasingly important for monitoring patient progression. However, such datasets are often small, incomplete, or have inconsistencies between observations. Thus, we propose a generative model that not only produces continuous trajectories of fully synthetic patient images, but also imputes missing data in existing trajectories, by estimating realistic progression over time. Our generative model is trained directly on features extracted from images and maps these into a linear trajectory in a Euclidean space defined with velocity, delay, and spatial parameters that are learned directly from the data. We evaluated our method on toy data and face images, both showing simulated trajectories mimicking progression in longitudinal data. Furthermore, we applied the proposed model on a complex neuroimaging database extracted from ADNI. All datasets show that the model is able to learn overall (disease) progression over time.

Keywords: Longitudinal data · Generative model · Synthetic images

1 Introduction

Longitudinal medical image data are important for *e.g.* modelling disease progression [1,27] or monitoring treatment response [3]. However, such datasets often suffer from incomplete or inconsistent observations, and are often limited in terms of size, diversity, and balance. Generally, using inadequate data can lead to

C. Chadebec and E. M. C. Huijben—Equal contribution.
S. Allassonnière and M. A. J. M. van Eijnatten—Equal contribution.

Supplementary Information The online version contains supplementary material available at https://doi.org/10.1007/978-3-031-18576-2_6.

A. Mukhopadhyay et al. (Eds.): DGM4MICCAI 2022, LNCS 13609, pp. 55–64, 2022.
https://doi.org/10.1007/978-3-031-18576-2_6

poor performances when being used to train machine learning (ML) models [23] for medical image analysis tasks such as classification [26] or segmentation [15].

To increase the size and variability of (non-longitudinal) medical imaging datasets, conventional data augmentation techniques such as rotation, cropping, or more resourceful augmentations [11] have been widely used [25]. However, the improved performances of deep generative models have given them the potential to perform image synthesis. Examples of such models are Generative Adversarial Networks (GANs) [10], which generate realistic images using a discriminator that distinguishes between real and synthetic images, and Variational Autoencoders (VAEs) [14], which constrain image features to follow a given prior distribution in order to generate synthetic images. These models have shown potential for synthesizing medical images of various modalities such as magnetic resonance imaging (MRI) [5,6,24], computed tomography (CT) [8,22], X-ray [18,21], or positron emission tomography (PET) [2]. In addition, several methods have been proposed to address data imputation or progression modelling in longitudinal imaging data of *e.g.* MRI [13,17] or simulated discrete progressions [20].

Although the topics of inter- and extrapolating longitudinal (medical) imaging data are well studied, to the best of our knowledge there is no model that addresses both of these aspects at once and is able to continuously generate realistic trajectories. In this paper, we propose a new deep generative model that is capable of: (1) generating realistic progression in images, (2) imputing missing data in existing patient trajectories, and (3) producing synthetic images with corresponding trajectories of non-existent patients[1].

2 Proposed Method

We propose a new generative model for longitudinal imaging data that consists of two steps. In the first step, relevant features are extracted from the input images using a VAE, and the second step maps these features into a linear trajectory to account for the progression over time. In the following, we refer to an observation, *e.g.* an image, as $y_{i,j} \in \mathcal{Y}$, with $i \in [1, N]$ the individual's identifier, $t_j \in \mathbb{R}_+^*$, where $j \in [0, P_i]$ the time of the observation. N is the number of individuals and P_i is the number of observations of i after the first time visit t_0.

2.1 Feature Extraction

Medical images are often complex and high-dimensional data. Therefore, instead of proposing a model directly acting on images, we propose to first extract meaningful features using a VAE (referred to as *the VAE* in the following). We use an autoencoder because it constrains comparable images to be encoded into similar locations such that minor variations in the latent space lead to smooth transformations in the image space. Since we expect smooth progressions, the VAE is likely to directly unveil trajectories in the latent space, thereby facilitating the second step of our method (referred to as *the generative model* in the following). In the following $x_{i,j} \in \mathcal{M}$ refers to the features of observation $y_{i,j}$.

[1] Code and dataset details are available at https://github.com/evihuijben/longVAE.

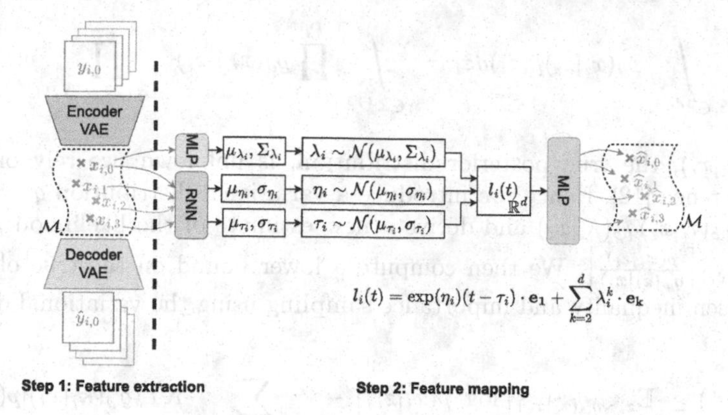

Fig. 1. Model sketch. First, features are extracted from images using the VAE (step 1), then, the proposed generative model maps these features to a straight line in Euclidean space (step 2). Network details are provided in Appendix 2.

2.2 Trajectory Modelling

We propose to learn parametric functions that map the features onto a linear trajectory in a d-dimensional Euclidean space $\mathbb{R}^d$ with standard basis $\{\mathbf{e_1}, \ldots, \mathbf{e_d}\}$, accounting for an individual's progression. We use the framework proposed in [17], in which an individual's progression trajectory at time t is modelled in $\mathbb{R}^d$ as

$$l_i(t) = \exp(\eta_i)(t - \tau_i) \cdot \mathbf{e_1} + \sum_{k=2}^{d} \lambda_i^k \cdot \mathbf{e_k}, \tag{1}$$

where η_i is a velocity parameter, τ_i is a delay, and $\lambda_i = (\lambda_i^k)_{2 \leq k \leq d}$ are spatial parameters. Contrary to [17], we adopt a fully variational approach to make the model generative in a similar fashion as [14]. Assuming a set of embeddings $x = \{(x_{i,j})_{1 \leq i \leq N, 0 \leq j \leq P_i}\} \in \mathcal{M}$, we first assume that given two individuals i and i', the features $x_{i,j}$ and $x_{i',j}$ are independent. Therefore, we propose to maximise the following likelihood objective $p(x) = \prod_{i=1}^{N} p(x_i)$, where $x_i = (x_{i,0}, \cdots, x_{i,P_i})$. We further assume that the latent variables $z_i = (\eta_i, \tau_i, \lambda_i^2, \cdots, \lambda_i^d) \in \mathbb{R}^{d+1}$ in Eq. (1) are such that the features of individual i at time t_j are generated by:

$$p_\theta(x_{i,j}|z_i) = \mathcal{N}\Big(\mu_\theta(l_i(t_j)), \sigma \cdot I_d\Big), \tag{2}$$

where $l_i(t_j)$ is the linear trajectory evaluated at t_j, and $\mu_\theta : \mathbb{R}^d \to \mathcal{M}$ is parameterised using a multilayer perceptron (MLP) and maps $\mathbb{R}^d$ to the feature space. The variation introduced by the stochastic model is the d-dimensional unit matrix I_d multiplied by a positive constant σ. We further assume that η_i, τ_i, and λ_i are independent and that for a given individual i, the features $x_{i,j}$ taken conditionally to z_i are independent. Furthermore, prior distributions over the latent variables are: $\eta_i \sim \mathcal{N}(0, \sigma_\eta)$, $\tau_i \sim \mathcal{N}(0, \sigma_\tau)$, $\lambda_i \sim \mathcal{N}(0, I_{d-1})$, with the dataset dependent priors σ_η and σ_τ. Finally, the likelihood for an individual i is:

$$p(x_i) = \int_{z_i \in \mathbb{R}^{d+1}} p_\theta(x_i|z_i)p(z_i)dz_i = \int_{z_i \in \mathbb{R}^{d+1}} \prod_{j=0}^{P_i} p_\theta(x_{i,j}|z_i) \prod_{\kappa_i \in \{\eta_i, \tau_i, \lambda_i\}} p(\kappa_i)d\kappa_i. \tag{3}$$

Since $p(z_i|x_i)$, the true posterior distribution, is unknown, we rely on variational inference [12]. Hence, we introduce a variational distribution $q_\varphi(z_i|x_i) = q_\varphi(\eta_i|x_i)q_\varphi(\tau_i|x_i)q_\varphi(\lambda_i|x_i)$ and derive a new estimate of the likelihood $p(x_i) = \mathbb{E}_{z_i \sim q_\varphi(z_i|x_i)}\left[\frac{p(x_i,z_i)}{q_\varphi(z_i|x_i)}\right]$. We then compute a lower bound on the true objective using Jensen inequality and importance sampling using the variational distribution.

$$\log\ p(x_i) \geq \mathbb{E}_{z_i \sim q_\varphi(z_i|x_i)}\Big[\log\ p(x_i|z_i)\Big] - \sum_{\kappa_i \in \{\eta_i, \tau_i, \lambda_i\}} KL(q_\varphi(\kappa_i|x_i)|p(\kappa_i)), \tag{4}$$

with KL the Kullback-Leibler divergence. In practice, we use multivariate Gaussians as variational distributions: $\eta_i \sim \mathcal{N}(\mu_\varphi^{\eta_i}, \Sigma_\varphi^{\eta_i})$, $\tau_i \sim \mathcal{N}(\mu_\varphi^{\tau_i}, \Sigma_\varphi^{\tau_i})$ and $\lambda_i \sim \mathcal{N}(\mu_\varphi^{\lambda_i}, \Sigma_\varphi^{\lambda_i})$. The parameters for progression, η_i and τ_i, are estimated from an input sequence using a recurrent neural network (RNN), while the spatial parameters, $(\lambda_i^2, \ldots, \lambda_i^d)$, are computed from the features of the image acquired at time t_0, which are estimated by the first MLP. The implementation details of the RNN and MLP can be found in Appendix 2, and a sketch of the model is presented in Fig. 1. Taking only the first image's features for the spatial parameters allows to estimate their value even if only one observation is available and to generate possible future progressions. Finally, we obtain the following loss function for one individual (removing constant terms):

$$\mathcal{L}_i = \sum_{j=0}^{P_i} \|x_{i,j} - \mu_\theta(l_i(t_j))\|^2 + \sum_{\kappa_i \in \{\eta_i, \tau_i, \lambda_i\}} KL(q_\varphi(\kappa_i|x_i)|p(\kappa_i)). \tag{5}$$

After training, we can either 1) generate fully synthetic trajectories using the aforementioned prior distributions, 2) produce possible progressions for a given individual i by estimating its λ_i and varying η_i and τ_i, or 3) interpolate and extrapolate existing trajectories by estimating the latent variables. Image sequences are then generated by recovering the features corresponding to a linear trajectory evaluated at a given time using a second MLP and passing them to the decoder of the VAE. In practice, we sample λ_i with a mixture of Gaussians since [9] recently showed that this approach alleviates the low expressiveness of the prior and allows to generate more convincing samples.

3 Data

We evaluate the proposed model using three longitudinal datasets. The first dataset is a toy dataset referred to as Starmen[2] [7] consisting of 64×64 binary

[2] Downloaded from https://doi.org/10.5281/zenodo.5081988.

images of 1,000 individuals that portray synthetic transformations based on the longitudinal model of [4], captured in 10 observations per individual. The second dataset, CelebA (aligned and cropped version downloaded in 2021) [16], consists of 64×64 RGB images of celebrities' faces. To resemble longitudinal medical images, we converted these images to grayscale and applied a simulated progression model by applying a non-linear intensity transform, a growth factor, a rotation, and adding Gaussian noise. This dataset can be considered very challenging since the images undergo global and local geometric transformations and photometric variations. The last dataset was obtained from the Alzheimer's Disease Neuroimaging Initiative[3]. We used a total of 8,318 MRI scans, obtained from 1,799 subjects, with an average of 4.6 ± 2.3 scans per person. The average time between the first and the last scan was 2.9 ± 2.4 years. We selected the 100th axial slice of every preprocessed scan and cropped it to 182×182. The subject's ages were used to define the observation times for the generative model, which were normalised between the overall oldest and youngest age. Details of the datasets (*e.g.* preprocessing steps, progression model, data splits, and example image trajectories) can be found in Appendix 1.

4 Experiments

Most experiments in this section are performed using Starmen and CelebA because these datasets are fully controlled and allow visual evaluation by non-medical experts. ADNI is used to show that results can be extended to medical data. In what follows, the models are selected on the validation set and tested on an hold-out test set. Experimental and implementation details are provided in Appendix 2.

Feature Extraction and Reconstruction. First, we train the VAE on each training set, disregarding the longitudinal component, and confirm the hypothesis that the features directly unveil clear trajectories over time, as can be seen in Fig. 1b in Appendix 1. To justify that mapping those trajectories to linear ones (step 2 in Fig. 1) is not too constraining, we analyse the reconstruction results obtained by 1) only encoding and decoding test images using the VAE (*base*), and 2) training the generative model to map the extracted feature trajectories to straight lines (Eq. (1)), evaluate $l(t)$ at observation times and pass the corresponding features to the decoder of the VAE (*ours*). Figure 2a and 2c show the mean squared error (MSE) and structural similarity (SSIM), respectively, of the test set reconstructions. Note that the results obtained using the proposed model is not expected to be better than the one obtained using the

[3] Data used in preparation of this article were obtained from the Alzheimer's Disease Neuroimaging Initiative (ADNI) database (http://adni.loni.usc.edu). As such, the investigators within the ADNI contributed to the design and implementation of ADNI and/or provided data but did not participate in analysis or writing of this report. A complete listing of ADNI investigators can be found at: http://adni.loni.usc.edu/wp-content/uploads/how_to_apply/ADNI_Acknowledgement_List.pdf.

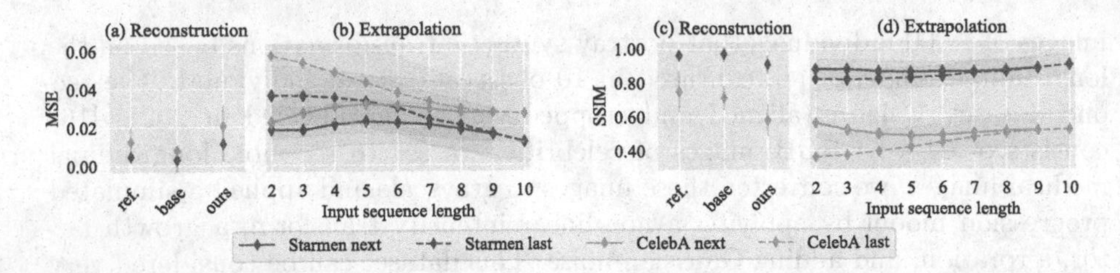

Fig. 2. Mean and standard deviation of MSE/SSIM (a, b/c, d) for various evaluations. (a/c) Metric between consecutive images in the test sequences (*ref.*) and reconstruction metrics using only the VAE (*base*) or the generative model (*ours*). (b/d) Metrics for the next and last image extrapolated based on a varying input sequence length.

VAE (*base*) because the generative model only acts on the features and we do not use any image-based reconstruction cost during its training. The metric values can be put into perspective by considering the mean value between two consecutive images in the test set (*ref.*). The visual reconstructions in the second row of Fig. 3 show that linear trajectory modelling does not considerably affect the image reconstruction ability of the model.

Fig. 3. Extrapolation of different test input sequences for Starmen (left) and CelebA (right). The first two rows represent the ground truth and reconstructions (*ours*), respectively. Red squares highlight images that were not provided to the model. Deviation from the true test Starmen image is presented in colour. (Color figure online)

Trajectory Extrapolation. In this section, we investigate whether the proposed model is able to extrapolate realistic trajectories from existing input data. To do so, we use the same model as before, but only provide the model with an image sequence of varying length and assess its ability to reconstruct either the next or the last image in the sequence. Figure 2b and Fig. 2d show the MSE and SSIM, respectively, of the ground truth and the extrapolated images based on a varying input sequence length. It can be seen that extrapolations become more reliable when a longer input sequence is given. This can also be observed from the visuals in Fig. 3, which show larger deviations from the ground truth when fewer images are presented. This experiment shows that in each case the model is able to estimate the progression: the left arm of the Starmen is raising and the CelebA head rotates, becomes bigger and contrast changes as expected. However, the model seems to underestimate the trajectory velocity as the input sequence becomes shorter. This aspect could potentially be mitigated by training using sequences of different lengths.

Data Imputation. We validate the ability of the model to impute missing data using input sequences simulating partial patient follow-ups. We simulate this by removing 50% of the training, validation, and test data acquired after t_0 using the Starmen and CelebA datasets. The VAE is trained using the 50% available images, after which the generative model learns to map the features onto a linear trajectory. In Fig. 4 we show the reconstructed samples at observation times.

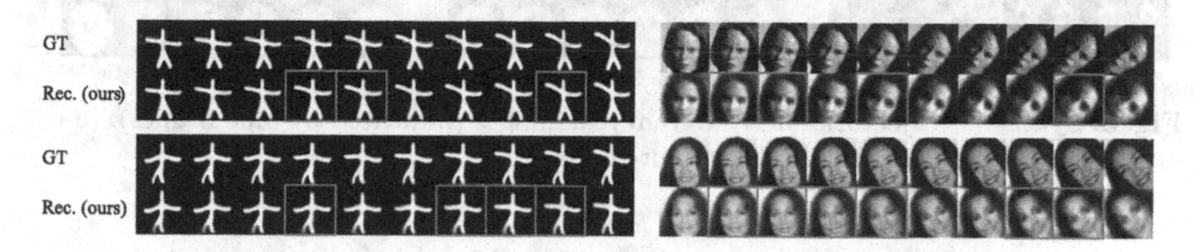

Fig. 4. Data imputation in test sequences with 50% missing data after t_0. Top rows show ground truth trajectories, red squares represent imputed images.(Color figure online)

Trajectory Generation. We also demonstrate that the proposed model can generate synthetic trajectories. We consider two cases: generating possible trajectories for a single image acquired at t_0 and generating a fully synthetic trajectory based on a synthetic image at t_0. In the first case, we first recover λ_i by encoding the real image using the VAE, estimate its value using the generative model and then sample η and τ from their priors as described in Sect. 2.2 and Appendix 2. In the second case, we first generate a synthetic λ and sample η and τ as aforementioned. To demonstrate the differences in these parameters, Fig. 5 shows trajectories obtained with varying delay τ (a) and velocity η (b), possible trajectories from an input image (c) and fully synthetic trajectories (d).

Real images are extracted from the test set and highlighted with blue frames. The results show that the proposed model allows to decorrelate spatial (λ) and time parameters (η and τ) since all images in a trajectory represent the same individual that undergoes smooth progressive change.

Neuroimaging Data. Finally, we validate the ability of the model to generate Alzheimer's disease progression trajectories. Figure 5e and 5f show trajectories generated from an existing input image and a synthetic image, respectively. The generated trajectories appear realistic because the ventricles grow over time, which is a marker of ageing and Alzheimer's disease progression [19]. Moreover, the proposed model seems to preserve the morphology represented at the first time point for both real and fake subjects. However, the generated disease progression trajectories still need to be assessed in more detail, for example by means of visual analysis by a medical expert or by training a deep learning-based classifier. Beside generating synthetic trajectories, we also investigate the extrapolation capability of the proposed model for the ADNI data, which is shown in

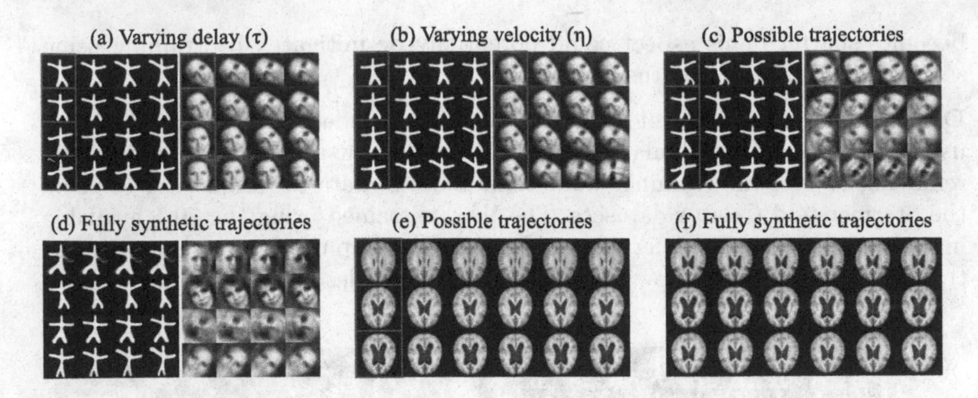

Fig. 5. Synthetic trajectories derived from real images (indicated by blue frames): (a-c, e) or synthetic images (d, f). (Color figure online)

Fig. 6. Contrary to the Starmen and CelebA experiments, this experiment shows a better performance for a shorter input sequence length. Generally, the proposed model seems to underestimate the disease progression (as estimated by η and τ), leading to a worse quantitative result for a later extrapolated sample.

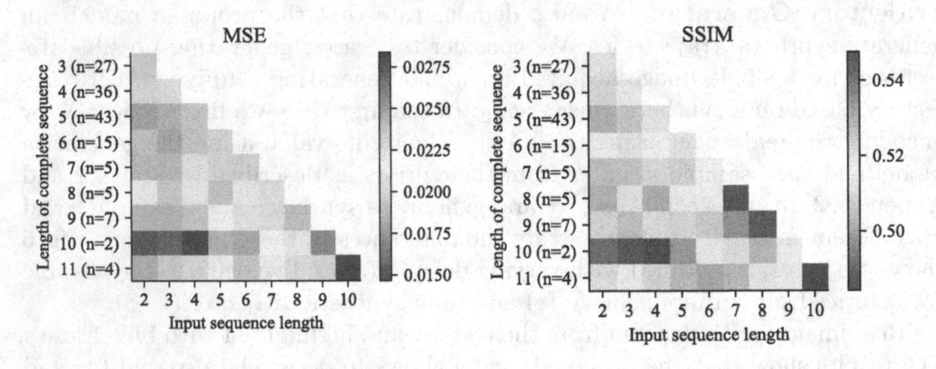

Fig. 6. Mean MSE (left) and SSIM (right) for the extrapolated next image after a given input sequence of the ADNI test set, with n the number of subjects. For interpretation of the colour bars, the reader is referred to the online version. (Color figure online)

5 Discussion and Conclusion

In this study we proposed a new continuous generative model capable of synthesising longitudinal imaging data to perform trajectory extrapolation, data imputation and smooth and probable synthetic trajectory generation. A notable strength of our model lies in its *two-step* architecture, which allows substituting the VAE to make the model suitable for any data type, *e.g.* using clinical scores directly as features. We believe that this work is a step towards synthesis and augmentation of longitudinal medical (image) datasets. However, the model needs more optimisation for such a high-dimensional complex medical imaging dataset, and a better trade-off between dimensionality reduction and efficient training of the generative model should be investigated. Furthermore, the hypothesis of smooth trajectories could be put into perspective by considering the disentangled 'brain age' instead of the real patient's age [27]. Future work should also focus on validating the ability of the model to perform reliable data augmentation for ML-based classification tasks or assess its relevance to perform treatment response analysis.

References

1. Aghili, M., Tabarestani, S., Adjouadi, M., Adeli, E.: Predictive modeling of longitudinal data for Alzheimer's disease diagnosis using RNNs. In: Rekik, I., Unal, G., Adeli, E., Park, S.H. (eds.) PRIME 2018. LNCS, vol. 11121, pp. 112–119. Springer, Cham (2018). https://doi.org/10.1007/978-3-030-00320-3_14
2. Bi, L., Kim, J., Kumar, A., Feng, D., Fulham, M.: Synthesis of positron emission tomography (PET) images via multi-channel generative adversarial networks (GANs). In: Cardoso, M.J., et al. (eds.) CMMI/SWITCH/RAMBO 2017. LNCS, vol. 10555, pp. 43–51. Springer, Cham (2017). https://doi.org/10.1007/978-3-319-67564-0_5
3. Blackledge, M.D., et al.: Assessment of treatment response by total tumor volume and global apparent diffusion coefficient using diffusion-weighted MRI in patients with metastatic bone disease: a feasibility study. PLoS ONE **9**(4), e91779 (2014)
4. Bône, A., Colliot, O., Durrleman, S.: Learning distributions of shape trajectories from longitudinal datasets: a hierarchical model on a manifold of diffeomorphisms. In: Proceedings of the IEEE Conference on Computer Vision and Pattern Recognition (CVPR), pp. 9271–9280 (2018)
5. Calimeri, F., Marzullo, A., Stamile, C., Terracina, G.: Biomedical data augmentation using generative adversarial neural networks. In: Lintas, A., Rovetta, S., Verschure, P.F.M.J., Villa, A.E.P. (eds.) ICANN 2017. LNCS, vol. 10614, pp. 626–634. Springer, Cham (2017). https://doi.org/10.1007/978-3-319-68612-7_71
6. Chadebec, C., Thibeau-Sutre, E., Burgos, N., Allassonnière, S.: Data augmentation in high dimensional low sample size setting using a geometry-based variational autoencoder. IEEE Trans. Pattern Anal. Mach. Intell. (2022)
7. Couronné, R., Vernhet, P., Durrleman, S.: Longitudinal self-supervision to disentangle inter-patient variability from disease progression. In: de Bruijne, M., et al. (eds.) MICCAI 2021. LNCS, vol. 12902, pp. 231–241. Springer, Cham (2021). https://doi.org/10.1007/978-3-030-87196-3_22
8. Frid-Adar, M., Diamant, I., Klang, E., Amitai, M., Goldberger, J., Greenspan, H.: GAN-based synthetic medical image augmentation for increased CNN performance in liver lesion classification. Neurocomputing **321**, 321–331 (2018)
9. Ghosh, P., Sajjadi, M.S., Vergari, A., Black, M., Schölkopf, B.: From variational to deterministic autoencoders. In: International Conference on Learning Representations (ICLR) (2020)
10. Goodfellow, I., et al.: Generative adversarial nets. In: Advances in Neural Information Processing Systems, vol. 27 (2014)
11. Hussain, Z., Gimenez, F., Yi, D., Rubin, D.: Differential data augmentation techniques for medical imaging classification tasks. In: AMIA Annual Symposium Proceedings, vol. 2017, p. 979. American Medical Informatics Association (2017)

12. Jordan, M.I., Ghahramani, Z., Jaakkola, T.S., Saul, L.K.: An introduction to variational methods for graphical models. In: Jordan, M.I. (ed.) Machine Learning. NATO ASI Series, pp. 105–161. Springer, Dordrecht (1998). https://doi.org/10.1007/978-94-011-5014-9_5
13. Kim, S.T., Küçükaslan, U., Navab, N.: Longitudinal brain MR image modeling using personalized memory for Alzheimer's disease. IEEE Access **9**, 143212–143221 (2021)
14. Kingma, D.P., Welling, M.: Auto-encoding variational bayes. arXiv preprint arXiv:1312.6114 (2013)
15. Liu, X., Song, L., Liu, S., Zhang, Y.: A review of deep-learning-based medical image segmentation methods. Sustainability **13**(3), 1224 (2021)
16. Liu, Z., Luo, P., Wang, X., Tang, X.: Deep learning face attributes in the wild. In: Proceedings of the IEEE International Conference on Computer Vision (ICCV) (2015)
17. Louis, M., Couronné, R., Koval, I., Charlier, B., Durrleman, S.: Riemannian geometry learning for disease progression modelling. In: Chung, A.C.S., Gee, J.C., Yushkevich, P.A., Bao, S. (eds.) IPMI 2019. LNCS, vol. 11492, pp. 542–553. Springer, Cham (2019). https://doi.org/10.1007/978-3-030-20351-1_42
18. Madani, A., Moradi, M., Karargyris, A., Syeda-Mahmood, T.: Chest X-ray generation and data augmentation for cardiovascular abnormality classification. In: Medical Imaging 2018: Image Processing, vol. 10574, pp. 415–420. International Society for Optics and Photonics, SPIE (2018)
19. Nestor, S.M., et al.: Ventricular enlargement as a possible measure of Alzheimer's disease progression validated using the Alzheimer's disease neuroimaging initiative database. Brain **131**(9), 2443–2454 (2008)
20. Ramchandran, S., Tikhonov, G., Kujanpää, K., Koskinen, M., Lähdesmäki, H.: Longitudinal variational autoencoder. In: Proceedings of The 24th International Conference on Artificial Intelligence and Statistics. Proceedings of Machine Learning Research, vol. 130, pp. 3898–3906. PMLR (2021)
21. Salehinejad, H., Valaee, S., Dowdell, T., Colak, E., Barfett, J.: Generalization of deep neural networks for chest pathology classification in X-rays using generative adversarial networks. In: IEEE International Conference on Acoustics, Speech and Signal Processing (ICASSP), pp. 990–994 (2018)
22. Sandfort, V., Yan, K., Pickhardt, P.J., Summers, R.M.: Data augmentation using generative adversarial networks (CycleGAN) to improve generalizability in CT segmentation tasks. Sci. Rep. **9**(1), 16884 (2019)
23. Shin, H.C., et al.: Deep convolutional neural networks for computer-aided detection: CNN architectures, dataset characteristics and transfer learning. IEEE Trans. Med. Imaging **35**(5), 1285–1298 (2016)
24. Shin, H.-C., et al.: Medical image synthesis for data augmentation and anonymization using generative adversarial networks. In: Gooya, A., Goksel, O., Oguz, I., Burgos, N. (eds.) SASHIMI 2018. LNCS, vol. 11037, pp. 1–11. Springer, Cham (2018). https://doi.org/10.1007/978-3-030-00536-8_1
25. Shorten, C., Khoshgoftaar, T.M.: A survey on image data augmentation for deep learning. J. Big Data **6**(1), 60 (2019)
26. Wen, J., et al.: Convolutional neural networks for classification of Alzheimer's disease: overview and reproducible evaluation. Med. Image Anal. **63**, 101694 (2020)
27. Zhao, Q., Liu, Z., Adeli, E., Pohl, K.M.: Longitudinal self-supervised learning. Med. Image Anal. **71**, 102051 (2021)

Applications

Novel View Synthesis for Surgical Recording

Mana Masuda[1](✉), Hideo Saito[1], Yoshifumi Takatsume[2], and Hiroki Kajita[2]

[1] Keio University, Yokohama, Kanagawa, Japan
{mana.smile,hs}@keio.jp

[2] Keio University School of Medicine, Shinjuku-ku, Tokyo, Japan
{tsume,jmrbx767}@keio.jp

Abstract. Recording surgery in operating rooms is one of the essential tasks for education and evaluation of medical treatment. However, recording the fields which depict the surgery is difficult because the targets are heavily occluded during surgery by the heads or hands of doctors or nurses. We use a recording system which multiple cameras embedded in the surgical lamp, assuming that at least one camera is recording the target without occlusion. In this paper, we propose Conditional-BARF (C-BARF) to generate occlusion-free images by synthesizing novel view images from the camera, aiming to generate videos with smooth camera pose transitions. To the best of our knowledge, this is the first work to tackle the problem of synthesizing a novel view image from multiple images for the surgery scene. We conduct experiments using an original dataset of three different types of surgeries. Our experiments show that we can successfully synthesize novel views from the images recorded by the multiple cameras embedded in the surgical lamp.

Keywords: Surgery recording · Generative model · Novel view synthesis

1 Introduction

Recording surgeries with cameras is indispensable for many reasons, such as education, sharing surgery technologies and techniques, performing case studies of diseases, and evaluating medical treatment [4,9,14,17]. Video recording is one of the simplest ways of recording surgery, and various methods have been proposed to record surgery.

It is hard to record the field which depicts the surgery without occlusion. The simplest way to record surgery is to attach the camera to the operating room environment. It may occur in the surgical field occluded by the doctors, nurses, or surgical machines, and the camera attached to the operating room environment is not suitable for recording surgery. The other way to record surgery is to attach the camera to the head of the doctor and record from the first-person view.

A. Mukhopadhyay et al. (Eds.): DGM4MICCAI 2022, LNCS 13609, pp. 67–76, 2022.
https://doi.org/10.1007/978-3-031-18576-2_7

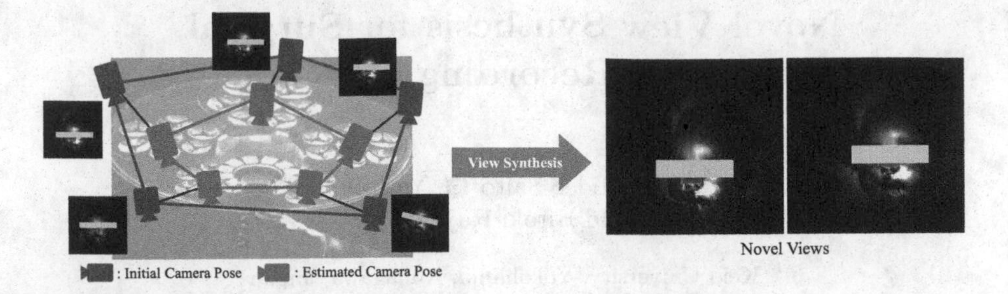

Fig. 1. An overview of our novel view synthesis framework. From the images given by the multiple cameras mounted on the surgical lamp, we conduct conditional bundle adjustment and train the neural radiance field (NeRF). Using the trained NeRF, we can synthesize novel view images.

This video is often affected by motion blur because of the head movements, and doctors do not always look at the surgical field. Therefore, a first-person viewpoint camera is also not suitable for recording surgery (Fig. 1).

Shimizu *et al.* [17] proposed a novel surgical lamp system with multiple embedded cameras to record surgeries. A generic surgical lamp has multiple light bulbs that illuminate the surgical field from multiple directions to reduce the shadows caused by the heads/hands of doctors or nurses. Using this system, [4,17] expected that at least one of the multiple light bulbs would always illuminate the surgical field. In the same way, they expected that at least one of the cameras embedded in the surgical lamp system always record the target without occlusion and proposed the method to select the best view recorded from these cameras. However, the problem with these methods is that the low quality of the video is due to frequent changes in the viewing direction, and it cannot be used effectively for its purpose, such as education.

Recent advancements in neural rendering such as Neural Radiance Fields (NeRF) [10] have recently gained a surge of interest within the computer vision community for their power to synthesize photorealistic novel views of real-world scenes. Given a set of images paired with a camera pose, NeRF learns the intensities of each pixel for a given camera pose. In addition to that, NeRF can estimate camera pose. Lin *et al.* [8] proposed BARF which realized to recover from imperfect camera poses and learn the NeRF representation simultaneously.

In this paper, we propose Conditional-BARF to synthesize novel view images of the surgical field. We suppose synthesizing novel views without occlusion makes it possible to generate a video in which the camera pose is smoothly shifted to a position where the surgical field is always visible. Since the limited number of cameras makes it difficult to accurately estimate camera pose using algorithms such as COLMAP [15,16], we consider recovering camera pose from imperfect camera poses as in BARF. In BARF, all cameras are optimized without any relevance. However, there are fixed spatial camera pose relationship for the cameras embedded in surgical lights. We propose Conditional-BARF (C-BARF) which makes use of this *condition* effectively to recover the camera positions.

As there is no dataset available to the public containing surgery recordings via multiple cameras, we record our dataset using the system proposed by Shimizu *et al.* [17]. The surgeries are recorded at our university's school of medicine. We demonstrate that C-BARF can synthesize even in surgical situations using three different types of surgery with five cameras attached to the surgical lamp.

In summary, our contributions are as follows;

- We tackled the task of novel view synthesis from multiple images of surgery. To the best of our knowledge, this is the first work to tackle this problem.
- We propose Conditional-BARF (C-BARF) suitable for synthesizing novel view images from limited images.
- We create a dataset of three different kinds of surgeries recorded with multiple cameras. We conduct experiments to show the effectiveness of our C-BARF for novel view synthesis of the surgical field. Please also refer to our accompanying video.

2 Related Work

2.1 Surgical Recording Systems

As doctors have a duty to teach their surgical skills to future generations, they need to record their surgery and generate videos for trainees. Moreover, it is widely recognized the usefulness of surgery recording in terms of reviewing. The surgery, such as laparoscopic surgery, which is performed through the endoscope camera can be easily recorded. However, the surgery that the doctor directly sees, such as open surgery, is difficult to record because the head or hands of the doctors or nurses and medical machines hide the important field of the surgeries.

Many attempts have been made to record surgery in the surgical field. A camera arm system is presented by Kumar *et al.* [6] with a camera mounted on the arm to record the surgery. The camera arm is set to a position that does not get occluded by the doctor and is often set to a position far from the surgical field and it is also cumbersome to position the camera according to the surgical situation and environment. Another system that mounts a camera on the surgical lamp is presented by Byrd *et al.* [1]. However, the view is occluded by the doctor's head or body and it is difficult to observe the surgical field with a single camera without any occlusion.

Other attempts also have been made to record surgery with a surgical field camera placed between the eyes of a doctor. The camera of such recording systems was not high resolution and did not produce good video quality which can be used for recording because of their limited hardware system [9,11]. In addition to that, that system is not comfortable for doctors because they are forced to perform surgery with interference by the surgical camera itself and its codes. Nair *et al.* [12] recorded surgery with a high-resolution camera (GoPro Hero4) on the doctor's head. The video is not good for training video because it is hardly blurred and does not shoot the surgical field at all times.

Multi Camera Recording. To solve these problems, Shimizu *et al.* [17] proposed a novel surgical lamp system with multiple embedded cameras assuming that the surgical field can be observed by one of the attached cameras because at least one of the multiple light bulbs always illuminates the surgical field. Shimizu *et al.* proposed the method to select the best view recorded from these cameras using Dijkstra's algorithm based on the area size of the surgery field. Hachiuma *et al.* [4] proposed Deep Selection which selects the camera with the best view of the surgery using a deep neural network in a fully supervised manner.

2.2 Novel View Synthesis

Novel view synthesis is one of the fundamental functionality and long-standing problem of computer vision. The first approach is simple light field sample interpolation techniques [2,3,7]. The computer vision and graphics communities have made great strides in predicting conventional geometric and appearance representations from observed images.

Neural Radiance Fields. Mildenhall *et al.* [10] proposed NeRF to synthesize novel views of static, complex scenes from a set of input images with known camera poses. The key idea of NeRF is to model the continuous radiance field of a scene with a multi-layer perceptron (MLP), followed by differentiable volumetric rendering to synthesize the images and backpropagate the photometric errors. In addition to that, some works have been made to realize a simultaneous camera pose tracking algorithm using NeRF [8,18,19]. iNeRF [19] proposed camera pose estimation algorithm using pre-trained NeRF. NeRF– [18] proposed two-stage pipeline to estimate unknown camera poses. Lin *et al.* [8] proposed BARF which is a simple coarse-to-fine bundle adjustment technique, we can recover from imperfect camera poses (including unknown poses of video sequences) and learn the NeRF representation simultaneously.

In our problem setup, we need to estimate the camera positions using camera pose estimation algorithms like COLMAP [15,16] to synthesize novel view images, but the limited number of cameras makes it difficult to estimate the exact position of cameras. On the other hand, the spatial camera pose relationship is known in advance from the camera setup settings (In our setting, the relationship is a regular polygon). We propose Conditional-BARF (C-BARF) which takes this advantage to perform bundle adjustment and synthesize novel view images.

3 Approach

3.1 Nural Radiance Fields

NeRF [10] encodes a 3D scene as a continuous 3D representation using MLP $\mathcal{G}_\Theta$ parameterized by learned weight Θ. The $\mathcal{G}_\Theta$ output the intensity $\mathbf{c}$ and volume density σ given a viewing ray $\mathbf{d}$ and a 3D coordinate $\mathbf{x}$. Given pixel coordinates

$\mathbf{u} \in \mathbb{R}^2$ and its homogeneous coordinate as $\bar{\mathbf{u}} = (x, y, 1)^T$, the 3D point $\mathbf{x}^i$ along the viewing ray $\mathbf{d}$ at depth z^i can be expressed as $\mathbf{x}^i = \bar{\mathbf{u}} + z^i\mathbf{d}$. From a 6DoF camera pose parametrized by $P \in \mathbb{R}^6$, we can estimate a 3D point $\mathbf{x}$ and viewing ray $\mathbf{d}$ through a 3D rigid transformation. The RGB color $\hat{I}$ at pixel location $\mathbf{u}$ is extracted by volume rendering via

$$\hat{I}(\mathbf{u}) = \int_{z_n}^{z_f} T(\mathbf{u}, z)\sigma(\bar{\mathbf{u}} + z\mathbf{d})\mathbf{c}(\bar{\mathbf{u}} + z\mathbf{d})dz, \tag{1}$$

where $T(\mathbf{u}, z) = \exp\left(-\int_{z_n}^{z} \sigma(\bar{\mathbf{u}} + z'\mathbf{d})dz'\right)$, z_n and z_f are bounds on the depth range of interest. In practice, this rendering function is approximated numerically via quadrature on points in a depth direction.

3.2 Conditional BARF

Our goal is to synthesize the novel view images from a set of pairs of image and camera pose $\mathbf{J} = \{\{I_1, P_1\}, \{I_2, P_2\}, \cdots, \{I_N, P_N\}\}$ where N is the number of cameras. Since the limited number of cameras makes it difficult to accurately estimate camera pose using algorithms such as COLMAP [15,16], we need to recover correct camera poses to synthesize clear novel view images. To achieve that goal, our problem is to optimize NeRF $\mathcal{G}_\Theta$ and the camera poses $\{P_i\}_{i=1}^N$ over the objective

$$\min_{P_1,\cdots,P_N,\Theta} \sum_{i=1}^{N} \sum_{\mathbf{u}} ||\hat{I}(\mathbf{u}; P_i, \Theta) - I_i(\mathbf{u})||_2^2. \tag{2}$$

Since the cameras are arranged in the shape of regular polygons, we can represent the spatial positional relationship between cameras as follows;

$$P_k = \mathbf{x}_c + \cos\left(\frac{2\pi}{N}k\right)\boldsymbol{a} + \sin\left(\frac{2\pi}{N}k\right)\boldsymbol{b}, \tag{3}$$

where $\mathbf{x}_c$ is the center of the cameras and $\boldsymbol{a} \cdot \boldsymbol{b} = 0$. We propose to optimize NeRF along with camera pose by solving Eq. (2) under the condition of camera setup settings (Eq. (3)). For optimization, we applied coarse-to-fine registration algorithm proposed in BARF [8].

4 Experiments

4.1 Dataset

As there is no dataset available that contains surgery recordings with multiple cameras, we use the system proposed by Simizu *et al.* [17] to create our dataset. The surgeries are recorded at Keio University School of Medicine. Video recording of the patients is approved by Keio University School of Medicine Ethics

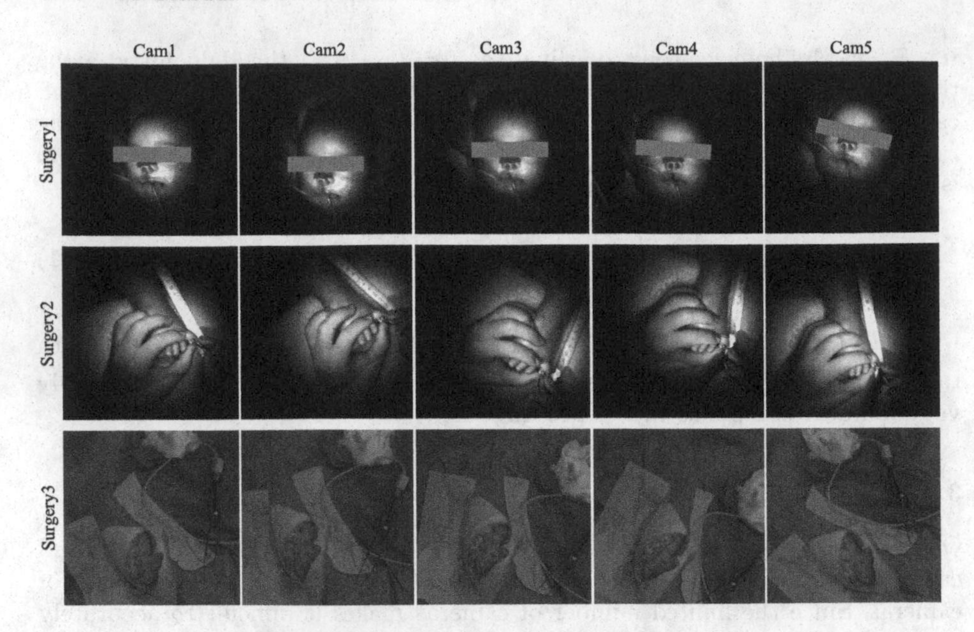

Fig. 2. Images recorded by the multiple cameras mounted on the surgical lamp. C-BARF experimented with three very different types of surgery (palatoschisis, polysyndactyly, and cicatrization). As in Shimizu *et al.* [17] and Hachiuma *et al.* [4], five cameras is attached to the surgical lamp. ($N = 5$). Faces are mosaicked for anonymity.

Committee and written informed consent s obtained from all patients or parents. We record three different types of surgery (cleft lip, polysyndactyly, and scar revision) with five cameras attached to the surgical lamp in a regular pentagon manner. We extract frames as shown in Fig. 2 from the videos at arbitrary timing and the initial camera position was estimated using COLMAP structure-from-motion package [15]. Our experiments are limited to qualitative evaluation due to the difficulty of generating test data, given the nature of the experiments using actual surgical scenes.

4.2 Experimental Settings

We optimize a separate neural continuous volume representation network for each surgery. We follow the same network architectural settings from the original NeRF [10] with some modifications as proposed in BARF [8]. We resized the images to 400×400 pixels. We train all models for 200K iterations and randomly sample 2048 pixel rays at each optimization step. We use the Adam optimizer [5] for both optimizing network and pose with a learning rate of 1×10^{-3} for the network decaying to 1×10^{-4}, and 3×10^{-3} for the pose P decaying to 1×10^{-5} following the original settings of BARF [8]. We used PyTorch [13] for implementation.

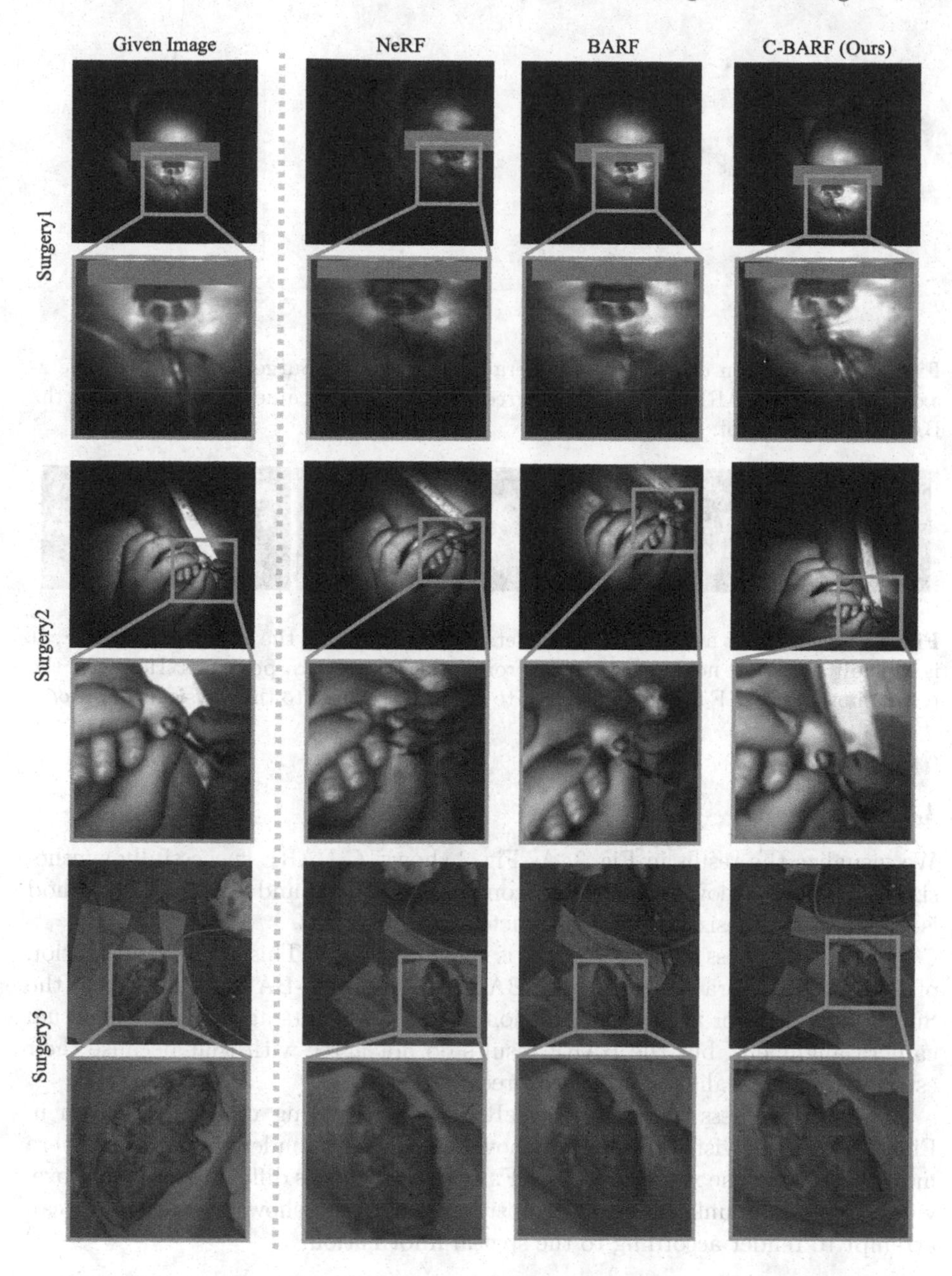

Fig. 3. Visual result of novel view images. Our proposed C-BARF successfully synthesizes high-fidelity novel view images. Compared to NeRF and BARF, C-BARF is better at synthesizing detailed structure and textures.

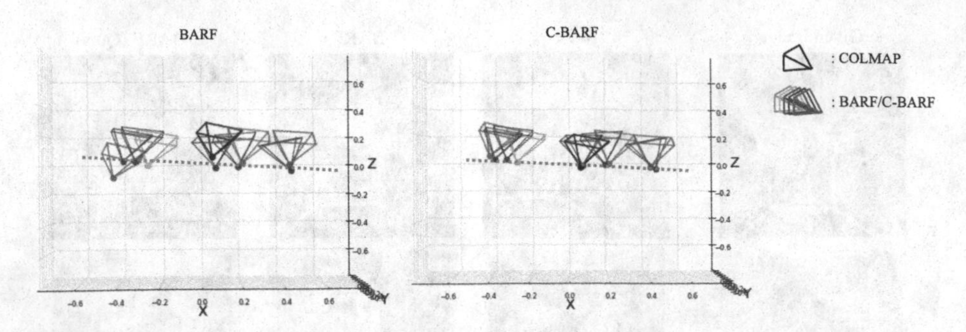

Fig. 4. Visualization of optimized camera poses from the Surgery1 scene. For the z-axis direction, C-BARF (ours) results agree with the actual camera alignment, but the BARF results do not.

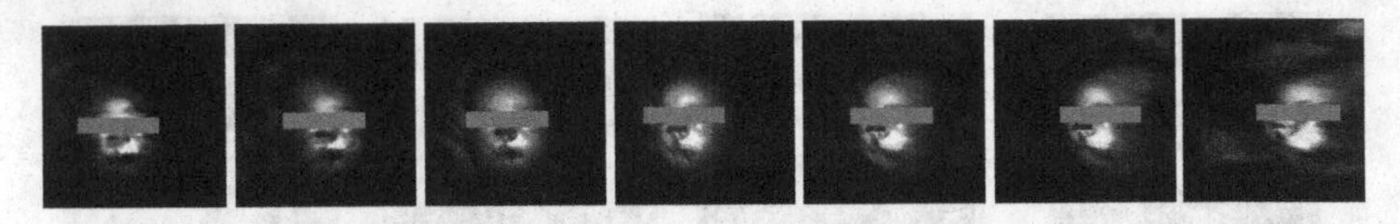

Fig. 5. Visual result along with the camera movement along the x-axis of C-BARF. It is difficult to render novel view images from unknown camera positions. However, the result shows C-BARF is able to attempt to render according to the physical conditions.

4.3 Result

We visualize the result in Fig. 3. As Fig. 3 shows, C-BARF successfully synthesizes high-fidelity novel view images compared to NeRF and BARF, our method is better at synthesizing detailed structure and textures.

The effectiveness of using Eq. (3) is shown in Fig. 4. This is the visualization of optimized camera poses for both BARF (left) and C-BARF (right) from the Surgery1 scene. For the z-axis direction, C-BARF results agree with the actual camera alignment, but the BARF results do not agree with that because their estimated camera alignment is scattered.

The effectiveness of using the NeRF-based rendering method is shown in Fig. 5. This is the visual result of the novel view images rendered with the camera movement along the x-axis of C-BARF. Even though it is difficult to render novel view images from unknown camera positions, the result shows that C-BARF can attempt to render according to the spatial information.

5 Conclusions

We tackled, for the first time, the task of novel view synthesis from multiple images of surgery. We proposed Conditional-BARF (C-BARF), a novel view synthesis method that is specialized for our novel surgical lamp system with multiple embedded cameras. The number of cameras is limited and the given camera poses are not accurate, but there is geometric prior information on the relative positions of the cameras. We took advantage of the geometric constraints for better synthesized novel views. Our experiments on our original datasets revealed that our method successfully makes use of the advantages and show better result compared to existing neural-network-based novel view synthesis methods.

Acknowledgement. We would like to express our gratitude to Yusuke Sekikawa, Denso IT Laboratory, Japan. Without his kind advice, this work would not have been completed. We also would like to thank the reviewers for their valuable comment. This work was supported by MHLW Health, Labour, and Welfare Sciences Research Grants Research on Medical ICT and Artificial Intelligence Program Grant Number 20AC1004, the MIC/SCOPE 201603003, and JSPS KAKENHI Grant Number 22H03617.

References

1. Byrd, R.J., Ujjin, V.M., Kongchan, S.S., Reed, H.D.: Surgical lighting system with integrated digital video camera. uS Patent 6,633,328, 14 October 2003
2. Davis, A., Levoy, M., Durand, F.: Unstructured light fields. In: Computer Graphics Forum, vol. 31, pp. 305–314. Wiley Online Library (2012)
3. Gortler, S.J., Grzeszczuk, R., Szeliski, R., Cohen, M.F.: The lumigraph. In: Proceedings of the 23rd Annual Conference on Computer Graphics and Interactive Techniques, pp. 43–54 (1996)
4. Hachiuma, R., Shimizu, T., Saito, H., Kajita, H., Takatsume, Y.: Deep selection: a fully supervised camera selection network for surgery recordings. In: Martel, A.L., et al. (eds.) MICCAI 2020. LNCS, vol. 12263, pp. 419–428. Springer, Cham (2020). https://doi.org/10.1007/978-3-030-59716-0_40
5. Kingma, D.P., Ba, J.: Adam: a method for stochastic optimization. arXiv preprint arXiv:1412.6980 (2014)
6. Kumar, A.S., Pal, H.: Digital video recording of cardiac surgical procedures. Ann. Thorac. Surg. **77**(3), 1063–1065 (2004)
7. Levoy, M., Hanrahan, P.: Light field rendering. In: Proceedings of the 23rd Annual Conference on Computer Graphics and Interactive Techniques, pp. 31–42 (1996)
8. Lin, C.H., Ma, W.C., Torralba, A., Lucey, S.: BARF: bundle-adjusting neural radiance fields. In: Proceedings of the IEEE/CVF International Conference on Computer Vision, pp. 5741–5751 (2021)
9. Matsumoto, S., et al.: Digital video recording in trauma surgery using commercially available equipment. Scand. J. Trauma Resuscitation Emerg. Med. **21**(1), 1–5 (2013)
10. Mildenhall, B., Srinivasan, P.P., Tancik, M., Barron, J.T., Ramamoorthi, R., Ng, R.: NeRF: representing scenes as neural radiance fields for view synthesis. In: Vedaldi, A., Bischof, H., Brox, T., Frahm, J.-M. (eds.) ECCV 2020. LNCS, vol. 12346, pp. 405–421. Springer, Cham (2020). https://doi.org/10.1007/978-3-030-58452-8_24
11. Murala, J.S., Singappuli, K., Swain, S.K., Nunn, G.R.: Digital video recording of congenital heart operations with "surgical eye". Ann. Thorac. Surg. **90**(4), 1377–1378 (2010)

12. Nair, A.G., et al.: Surgeon point-of-view recording: using a high-definition head-mounted video camera in the operating room. Indian J. Ophthalmol. **63**(10), 771 (2015)
13. Paszke, A., et al.: Automatic differentiation in pytorch. In: NIPS-W (2017)
14. Sadri, A., Hunt, D., Rhobaye, S., Juma, A.: Video recording of surgery to improve training in plastic surgery. J. Plast. Reconstr. Aesthetic Surg. **66**(4), e122–e123 (2013)
15. Schönberger, J.L., Frahm, J.M.: Structure-from-motion revisited. In: Conference on Computer Vision and Pattern Recognition (CVPR) (2016)
16. Schönberger, J.L., Zheng, E., Frahm, J.-M., Pollefeys, M.: Pixelwise view selection for unstructured multi-view stereo. In: Leibe, B., Matas, J., Sebe, N., Welling, M. (eds.) ECCV 2016. LNCS, vol. 9907, pp. 501–518. Springer, Cham (2016). https://doi.org/10.1007/978-3-319-46487-9_31
17. Shimizu, T., Oishi, K., Hachiuma, R., Kajita, H., Takatsume, Y., Saito, H.: Surgery recording without occlusions by multi-view surgical videos. In: VISIGRAPP (5: VISAPP), pp. 837–844 (2020)
18. Wang, Z., Wu, S., Xie, W., Chen, M., Prisacariu, V.A.: NeRF: neural radiance fields without known camera parameters. arXiv preprint arXiv:2102.07064 (2021)
19. Yen-Chen, L., Florence, P., Barron, J.T., Rodriguez, A., Isola, P., Lin, T.Y.: iNeRF: inverting neural radiance fields for pose estimation. In: IEEE/RSJ International Conference on Intelligent Robots and Systems (IROS) (2021)

Anomaly Detection Using Generative Models and Sum-Product Networks in Mammography Scans

Marc Dietrichstein[1], David Major[1(✉)], Martin Trapp[2], Maria Wimmer[1], Dimitrios Lenis[1], Philip Winter[1], Astrid Berg[1], Theresa Neubauer[1], and Katja Bühler[1]

[1] VRVis Zentrum für Virtual Reality und Visualisierung Forschungs-GmbH, Vienna, Austria
david.major@vrvis.at

[2] Department of Computer Science, Aalto University, Espoo, Finland

Abstract. Unsupervised anomaly detection models that are trained solely by healthy data, have gained importance in recent years, as the annotation of medical data is a tedious task. Autoencoders and generative adversarial networks are the standard anomaly detection methods that are utilized to learn the data distribution. However, they fall short when it comes to inference and evaluation of the likelihood of test samples. We propose a novel combination of generative models and a probabilistic graphical model. After encoding image samples by autoencoders, the distribution of data is modeled by Random and Tensorized Sum-Product Networks ensuring exact and efficient inference at test time. We evaluate different autoencoder architectures in combination with Random and Tensorized Sum-Product Networks on mammography images using patch-wise processing and observe superior performance over utilizing the models standalone and state-of-the-art in anomaly detection for medical data.

Keywords: Anomaly detection · Generative models · Sum-product networks · Mammography

1 Introduction

Acceleration of the detection and segmentation of anomalous tissue by automated computer-aided approaches is a key to enhancing cancer screening programs. It is especially important for mammography screening, as breast cancer is the most common cancer type and the leading cause of death in women worldwide [20]. Training an artificial neural network in a supervised way requires a high amount of pixel-wise annotated data. As data annotation is very costly, methods that involve as less annotation as possible are in high demand. Anomaly

M. Dietrichstein and D. Major—Equal contribution.

A. Mukhopadhyay et al. (Eds.): DGM4MICCAI 2022, LNCS 13609, pp. 77–86, 2022.
https://doi.org/10.1007/978-3-031-18576-2_8

detection approaches are good representatives of this type, as they only utilize healthy cases for learning, and anomalous spots are detected as a deviation from the learned data distribution. The deviation is measured either by straightforward metrics such as reconstruction error of input and output samples or by more sophisticated constructs such as log-likelihood in probabilistic models.

Unsupervised anomaly detection methods have been evaluated on a plethora of different pathologies and medical imaging modalities. A state-of-the-art method in this area is f-AnoGAN [15], which leverages Generative Adversarial Networks (GANs) to model an implicit distribution of healthy images and detect outliers via a custom anomaly score based on reconstruction performance. f-AnoGAN has been utilized to detect anomalies in Optical Coherence Tomography (OCT) scans [15], Chest X-rays [1], and 3D Brain scans [17]. However, it requires the training of a separate encoder module to obtain latent codes of images, which are used by the generator for reconstruction. The autoencoder (AE) architecture, on the other hand, jointly trains an encoder and decoder and is thus able to directly map an input to its corresponding latent representation. AE variants have been applied to lesion detection in mammography images [19] and brain scans [8,21], as well as head [14] and abdomen [8] Computed Tomography scans. However, the practical applicability of all those models is limited by the fact that the respective anomaly scores are not easily interpretable by a human decision maker. Here, to remedy the situation, it would be desirable for the model to provide some degree of certainty for its decision. To this end, density estimation models can be employed. Such models learn an explicit probability density function from the training data and assume that anomalous samples are located within low-density regions. Examples are the application of Gaussian Mixture Models [2] for brain lesion detection as well as Bayesian U-Nets for OCT anomaly detection [16]. Although these approaches are similar to ours, they are tailored to specific image modalities and can thus not be directly applied to our domain.

In this work, we introduce a novel and general method for anomaly detection that combines AEs with probabilistic graphical models called Sum-Product Networks (SPNs). A recent powerful SPN architecture called Random and Tensorized SPN (RAT-SPN) [12] was chosen, as it is easy to integrate into deep learning frameworks and is trained by GPU-based optimization. More than that, standard and variational AEs do not allow to derive exact data likelihoods, they rather provide approximations that can be used for anomaly detection. SPNs solve this problem and allow *exact* and *efficient* likelihood inference by imposing special structural constraints on the model capturing the data distribution. We compare the performance of different standalone AEs to that of their combination with RAT-SPNs on unsupervised mass and calcification detection in public mammography scans and demonstrate improvements.

2 Methods

Our approach learns the healthy data distribution in a patch-wise fashion. First, the dimensionality of patch data is reduced by an AE, and the likelihood for

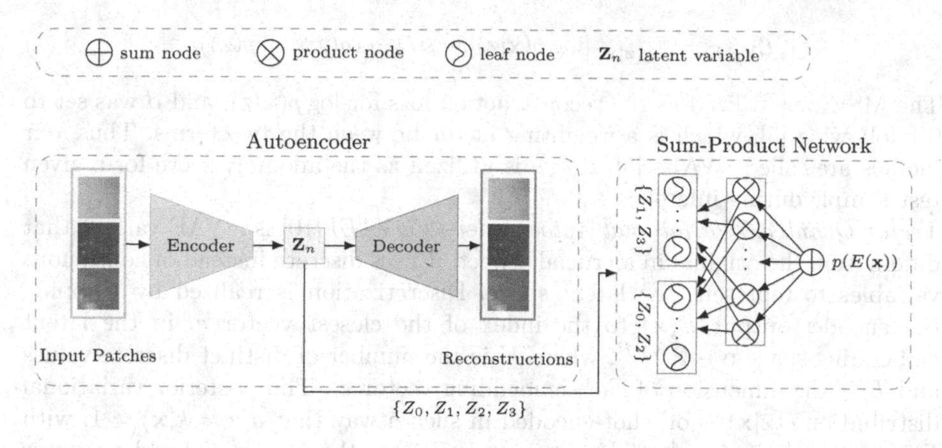

Fig. 1. The encoder of an AE outputs a low-dimensional latent representation $\mathbf{z}_n$ of healthy mammography patches. This representation serves then as input to a SPN that learns the corresponding probability distribution $p(E(\mathbf{x}))$. The likelihood of test samples is predicted over the same pipeline using trained models.

membership to the data distribution is approximated by a RAT-SPN. During inference, the learned model is applied to test images patch by patch, where at every position, the likelihood yields the anomaly score. As the models capture the distribution of healthy data, this score should be significantly different at anomalous image positions. We compare the performance of our pipeline to that of standalone AE models. The different AEs, that we considered, are described in Sect. 2.1 and RAT-SPNs in Sect. 2.2. It is followed by our proposed combination of an AE with a RAT-SPN in Sect. 2.3. A system overview is provided in Fig. 1.

2.1 Autoencoders

Convolutional AEs (CAEs) [9] utilize convolutional blocks to map high dimensional image data $\mathbf{x} \in \mathbb{R}^{H \times W}$ into a lower dimensional latent space $\mathbf{z} \in \mathbb{R}^M$ using an encoder by $\mathbf{z} = E(\mathbf{x})$ and reconstruct it utilizing a decoder model by $\hat{\mathbf{x}} = D(E(\mathbf{x}))$. The compression and reconstruction process is learned by minimizing the reconstruction loss $\mathcal{L}_{\text{CAE}} = \ell_2(\mathbf{x}, \hat{\mathbf{x}})$ where ℓ_2 signalizes the mean squared error (MSE). Computation of $\mathcal{L}_{\text{CAE}}$ for test samples yields the anomaly score at inference.

Variational Autoencoders (VAEs) [6] are equipped with the same building blocks as CAEs when applied to images, but additionally, they aim to approximate the true posterior distribution $p(\mathbf{z}|\mathbf{x})$ in the encoder E by a simpler and more tractable distribution $q(\mathbf{z}|\mathbf{x})$. This is achieved by minimizing the KL divergence $D_{KL}(q(\mathbf{z}|\mathbf{x}) \,||\, p(\mathbf{z}|\mathbf{x}))$ between the two distributions. On the other hand, the decoder D learns the posterior $p(\mathbf{x}|\mathbf{z})$ and reconstructs $\mathbf{x}$ from a given $\mathbf{z}$ by maximizing the log-likelihood $\log p(\mathbf{x}|\mathbf{z})$. The overall objective to minimize is called the evidence lower bound (ELBO), and it can be formulated as follows:

$$\mathcal{L}_{VAE} = \mathbb{E}_{q(\mathbf{z}|\mathbf{x})}\left[\log p(\mathbf{x}|\mathbf{z})\right] - \beta D_{KL}(q(\mathbf{z}|\mathbf{x}) \,||\, p(\mathbf{z})). \tag{1}$$

The MSE was utilized as the reconstruction loss for $\log p(\mathbf{x}|\mathbf{z})$, and β was set to 0.1 following [5], which is a weighting factor between the two terms. Thus, our models are called βVAEs [5]. $\mathcal{L}_{VAE}$ is utilized as the anomaly score for a given test sample during inference.

Vector Quantised-Variational Autoencoder (VQVAE) [10] is a VAE variant that differs from the original in a crucial aspect: it uses discrete instead of continuous variables to represent the latent space. Discretization is realized by mapping the encoder output $E(\mathbf{x})$ to the index of the closest vector e_i in the latent embedding space $\mathbf{e} \in \mathbb{R}^{K \times B}$, where K is the number of distinct discrete values and B is the dimension of each embedding vector e_i. The posterior variational distribution $q(\mathbf{z}|\mathbf{x})$ is one-hot-encoded in such a way that $q(\mathbf{z} = k|\mathbf{x}) = 1$, with $k = \arg\min_i ||E(\mathbf{x}) - e_i||_2$. The mapping of $E(\mathbf{x})$ to the nearest embedding vector e_i is defined as $E_q(\mathbf{x}) = e_k$ with $k = \arg\min_i ||E(\mathbf{x}) - e_i||_2$. The loss formulation consists of three parts, each aiming to optimize a different aspect of the model:

$$\mathcal{L}_{\text{VQVAE}} = \log p(\mathbf{x}|E_q(\mathbf{x})) + ||sg[E(\mathbf{x})] - \mathbf{e}||_2^2 + \lambda||E(\mathbf{x}) - sg[\mathbf{e}]||_2^2. \tag{2}$$

The first term is the reconstruction loss, for which MSE was again chosen. The remaining terms are concerned with learning an optimal embedding space. The codebook loss, the second term, attempts to move the embedding vectors closer to the encoder output, whereas the third term, the commitment loss, attempts the inverse and forces the encoder output to be closer to the closest embedding vector $\mathbf{e}$. $sg[\cdot]$ is the stop-gradient operator and prevents its operand from being updated during back-propagation. λ is a weighting factor for the commitment loss, which we set to 0.25, following [10]. The anomaly score for a given test sample is determined by calculating its reconstruction loss.

2.2 Sum-Product Networks

SPNs [13] are tractable probabilistic models of the family of probabilistic circuits [3] and allow various probabilistic queries to be computed efficiently and exactly. For consistency with recent works, we will introduce SPNs based on the formalism in [18]. An SPN on a set of random variables $\mathbf{Z} = \{Z_j\}_{j=1}^{J}$ is a tuple $(\mathcal{G}, \psi)$ consisting of a computational graph $\mathcal{G}$, which is a directed acyclic graph, and a scope function ψ mapping from the set of nodes in $\mathcal{G}$ to the set of all subsets of $\mathbf{Z}$ including $\mathbf{Z}$. The computational graph of an SPN is typically composed of sum nodes, product nodes, and leaf nodes. Sum nodes compute a weighted sum of their children, i.e., $(\mathrm{S}(\mathbf{z}) = \sum_{\mathrm{N} \in \mathrm{ch(S)}} \theta_{\mathrm{S,N}}\, \mathrm{N}(\mathbf{z}))$, product nodes compute a product of their children, i.e., $(\mathrm{P}(\mathbf{z}) = \prod_{\mathrm{N} \in \mathrm{ch(P)}} \mathrm{N}(\mathbf{z}))$, and leaf nodes are tractable multivariate or univariate probability distributions or indicator functions. The scope function assigns each node a scope (subset of $\mathbf{Z}$ or $\mathbf{Z}$) and ensures that the SPN fulfills certain structural properties, guaranteeing that specific probabilistic queries can be answered tractably. In this work, we will focus on SPNs that are *smooth* and *decomposable*, we refer to [3] for a detailed discussion. Moreover, we consider a representation of the SPN in the form of a random and tensorized region graph called RAT-SPNs [12] and employ the implementation based on Einstein summation as proposed in [11]. The region graph is parametrized by the number of root nodes C, input distributions I as well as the graph depth D, and the number of parallel SPN instances, or recursive splits, R. By choosing these parameters, RAT-SPNs with arbitrary complexity may be constructed. From a given region graph, it is possible to obtain the underlying SPN structure in terms of its computational graph and scope function exactly, for more details see [12,18]. A simple region graph with $C = 1$, $I = 2$, $D = 1$, and $R = 1$, and the underlying SPN is illustrated in Fig. 1. In a generative learning setting like ours, the optimal network parameters w are found by applying (stochastic) Expectation Maximization (EM) to maximize the log-likelihood LL of the training samples:

$$LL(w) = \frac{1}{N} \sum_{n=1}^{N} \log \mathrm{S}(\mathbf{z}_n). \tag{3}$$

2.3 Combining Autoencoders and Sum-Product Networks

We combine each AE type of Sect. 2.1 with a RAT-SPN by passing the learned latent representation of encoded samples $\mathbf{z} = E(\mathbf{x})$ as observed states for the random variables $\mathbf{Z}$ to a RAT-SPN (see Fig. 1). This way, after input images are mapped to a low-dimensional space, likelihoods can be obtained exactly and efficiently in an end-to-end fashion at inference. Extra computations, essential for reconstruction and ELBO-based scores of standalone AEs, are therefore not necessary. Two RAT-SPN setups are utilized, one with Gaussian input distributions for the continuous latent representations of CAEs and βVAEs, and the other with categorical inputs for the discrete features of VQVAEs. Training is done separately, first the AE models are trained followed by RAT-SPNs. The anomaly score is yielded by the likelihood of a trained AE and RAT-SPN combination for a given test sample.

3 Experimental Setup

3.1 Datasets

We train our models on the Digital Database for Screening Mammography (DDSM) [4], a collection of 2620 mammography exams, with each exam consisting of multiple images. The images in this dataset are categorized according to the type of diagnosis, either into *healthy* or into a *cancer* type (i.e., malignant, benign). As we want to learn a healthy model, we selected only the 695 healthy exams containing 2798 images for training purposes. From each image, 120 patches of 64×64 pixels (px), the resolution also used by [15], were extracted; half of these containing internal breast tissue, the other half were sampled along the breast contour. We evaluated all methods against a selection of cancerous mammograms from the Curated Breast Imaging Subset of

DDSM (CBIS-DDSM) [7], which provides improved annotations of masses and calcifications for images from DDSM. In order to filter out images with large-scale annotations, we set the restriction that the annotation mask area must be smaller than 4-times of our patch area for masses, and it should contain the whole calcifications. 79 scans were selected from the mass and 30 scans from the calcification test set that fulfilled these criteria. The healthy training images consisted of equally distributed dense and non-dense tissues, whereas the mass test cases had a ratio of 14%/86% and the calcification test samples a ratio of 40%/60% (dense/non-dense).

3.2 Training

The 2798 healthy images were split into 90% training and 10% validation images (with no patient overlap) for training both the AE (CAE, βVAE, VQVAE) and the RAT-SPN models. Following [9], all of our AE models had an architecture with 32-64-128 2D convolutional layers with 5×5 kernels and a stride of 2 in the encoder and 2D transposed convolutional layers in the decoder. The VQVAE model had additional 6 residual blocks with 128 filters, and the dimensionality of the embedding vector was 64. All models had 64 latent units and were trained with a batch size of 64. CAEs and βVAEs were trained for 100 epochs with a learning rate of 1e$-$5, whereas VQVAEs converged to an optimum after 20 epochs with a learning rate of 1e$-$4. The best-fit RAT-SPN parameters of $C = 1$, $I = 45$, $D = 1$, $R = 50$ were found utilizing the 10% validation images, possible values were taken from the supplement of [12]. The RAT-SPN setup was the same for all AE models, and it was trained by the EM algorithm for 50 epochs with a batch size of 64 and a learning rate of 1e$-$4.

3.3 Evaluation

We evaluated our methods on the 79 mass and 30 calcification test images. The anomaly score assignment was performed in a lower dimensional image space than the original resolution, and thus, patches were sampled around every 16th pixel per image. Only breast tissue pixels were considered using pre-segmentations of the breast area in every image. In order to show the anomaly scores' discriminative power between healthy and anomalous positions, we derived the Area Under the ROC Curve (AUC) in two ways, either considering all pixels from all test images at once (pixel-wise) or doing it for each image separately and calculating the average over all test samples additionally (image-wise). In order to measure the capability of the methods for detection of the anomalous regions, we apply the Hausdorff distance (H) image-wise to assess the pixel distance between masks generated by our models and the provided CBIS-DDSM ground-truth. It measures the maximum of the distances from any annotated point in one mask to the nearest point in the other mask, thus, the smaller it is, the closer the match between prediction and ground-truth.

Table 1. Anomaly detection results utilizing different models. Metrics are computed either over pixels or images. Next to AUC scores average Hausdorff (H) distances (px) between anomaly segmentations and ground-truth masks were computed. Segmentations are calculated after score thresholding by 99th-percentile. Statistically significantly better performance (based on image-wise AUCs) between standalone and RAT-SPN extended models are depicted in bold ($p < 0.01$).

Test data	Model	Pixel-wise	Image-wise	
		AUC	AUC	H
Masses	CAE	0.53	0.58 ± 0.25	30.80 ± 12.57
	CAE-RATSPN	**0.88**	**0.88 ± 0.10**	**30.10 ± 14.50**
	βVAE	0.80	0.83 ± 0.14	30.78 ± 12.79
	βVAE-RATSPN	0.88	0.88 ± 0.11	29.07 ± 14.57
	VQVAE	0.67	0.67 ± 0.19	33.37 ± 12.53
	VQVAE-RATSPN	**0.82**	**0.84 ± 0.13**	**32.21 ± 14.01**
	f-AnoGAN	0.86	0.85 ± 0.12	30.41 ± 11.91
Calcifications	CAE	0.65	0.77 ± 0.21	32.79 ± 10.82
	CAE-RATSPN	0.72	0.78 ± 0.16	33.03 ± 13.74
	βVAE	0.73	0.80 ± 0.17	29.73 ± 12.56
	βVAE-RATSPN	0.66	0.73 ± 0.17	30.72 ± 13.07
	VQVAE	0.69	0.79 ± 0.17	31.79 ± 12.66
	VQVAE-RATSPN	0.68	0.75 ± 0.19	33.34 ± 10.61
	f-AnoGAN	0.67	0.74 ± 0.20	34.95 ± 6.98

4 Results and Conclusion

We compare the anomaly detection performance of the three AE models in their standalone configuration as well as with a RAT-SPN extension. Additionally, we trained and evaluated a state-of-the-art f-AnoGAN model in its default configuration. The results are illustrated in Table 1 and Fig. 2.

For the mass test set, the overall best performing model was the βVAE-RATSPN with 0.88 pixel-wise and average image-wise AUCs, and an average H-distance of 29.07 px (see Fig. 2). Statistically significant superior image-wise AUC performances over standalone models were achieved by CAE-RATSPNs and VQVAE-RATSPNs. Except for VQVAE-RATSPN, all RAT-SPN extended models performed better than f-AnoGAN in terms of image-wise AUC, although there were no statistically significant differences (cf. Table 1). It is also visible in Table 1 that RAT-SPNs applied to continuous features yielded better results than the discrete version. Furthermore, it is depicted in Fig. 3 a) and b), that attaching RAT-SPN models to AE models facilitate a better discrimination between healthy and anomalous tissue by increasing the gap between their respective distributions.

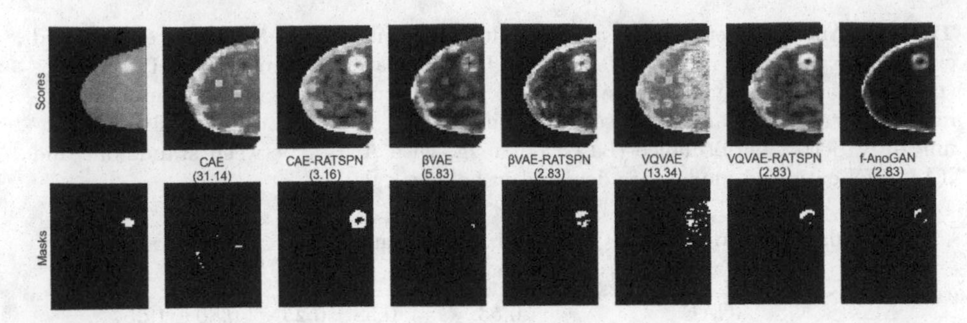

Fig. 2. Anomaly detection results of a mass sample. The first column shows the mammography scan (top) and anomaly ground-truth (bottom). The remaining columns depict an anomaly score heatmap (top) and a segmentation mask (bottom) for each model. Bright yellow pixels represent high and dark red pixels low anomaly scores in the heatmaps. The respective Hausdorff distance to the ground-truth (px) is displayed after each method name in brackets. Segmentations are calculated after score thresholding by 99th-percentile. (Color figure online)

Moreover, the standalone βVAE was the best performing model for the calcification test set with an 0.73 pixel-wise and 0.80 average image-wise AUC, and an average H-distance of 29.73 px. It is in general visible that all models reflect a consistently poorer performance for this data. This is due to the fact that this set contains a higher proportion of dense breasts than the mass collection (see Sect. 3.1), and most of the small calcifications were generally hard to detect accurately by all models in images dominated by dense tissue. On the other hand, the standalone versions performed here better than the ones with RAT-SPN extension except for the CAE setup, but no statistically significant differences were discovered based on the image-wise AUC scores (cf. Table 1). This behavior is well visualized by the score distribution plots of the best-performing standalone βVAE and βVAE-RATSPN versions in Fig. 3 c) and d). All models except for βVAE-RATSPN yielded better image-wise AUCs than f-AnoGAN, although no statistically significant differences were detected (cf. Table 1).

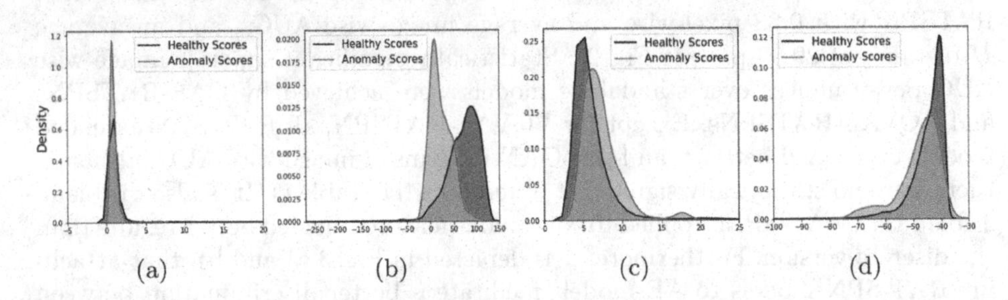

Fig. 3. Distribution of healthy and anomaly scores on the masses (a, b) and calcifications datasets (c, d) for CAE without (a) and with RAT-SPN extension (b), for βVAE without (c) and with RAT-SPN extension (d).

In summary, we have introduced a novel unsupervised anomaly detection method that extends various AE architectures with a RAT-SPN module. This approach is a promising avenue for generating exact likelihoods and incorporating them into the detection of different anomalies, such as masses and calcifications in mammography scans. Our experiments suggest that our method clearly outperforms standalone AE models on mass samples. Furthermore, it exhibits similar results to those of the state-of-the-art f-AnoGAN, however, with the advantages of a comparatively simpler training setup and exact likelihood inference. All of the investigated methods have difficulties when applied to calcification samples. We interpret that this is due to the presence of a larger proportion of dense tissue in the latter dataset. In future work, we plan to analyze how this problem can be eliminated and in particular whether increasing the input resolution has a positive effect on the performance.

Acknowledgement. VRVis is funded by BMK, BMDW, Styria, SFG, Tyrol and Vienna Business Agency in the scope of COMET - Competence Centers for Excellent Technologies (879730) which is managed by FFG. Thanks go to AGFA HealthCare, project partner of VRVis, for providing valuable input. Martin Trapp acknowledges funding from the Academy of Finland (347279).

References

1. Bhatt, N., Prados, D.R., Hodzic, N., Karanassios, C., Tizhoosh, H.R.: Unsupervised detection of lung nodules in chest radiography using generative adversarial networks. In: Proceedings of EMBC, pp. 3842–3845. IEEE (2021)
2. Bowles, C., et al.: Brain lesion segmentation through image synthesis and outlier detection. NeuroImage: Clin. **16**, 643–658 (2017)
3. Choi, Y., Vergari, A., Van den Broeck, G.: Probabilistic circuits: a unifying framework for tractable probabilistic models. Technical report, UCLA (2020)
4. Heath, M., Bowyer, K., Kopans, D., Moore, R., Kegelmeyer, W.P.: The digital database for screening mammography. In: Proceedings of the International Workshop on Digital Mammography, pp. 212–218. Medical Physics Publishing (2000)
5. Higgins, I., et al.: Beta-VAE: learning basic visual concepts with a constrained variational framework. In: ICLR (2017)
6. Kingma, D.P., Welling, M.: Auto-encoding variational Bayes. arXiv preprint arXiv:1312.6114 (2013)
7. Lee, R.S., Gimenez, F., Hoogi, A., Miyake, K.K., Gorovoy, M., Rubin, D.L.: A curated mammography data set for use in computer-aided detection and diagnosis research. Sci. Data **4**(1), 1–9 (2017)
8. Marimont, S.N., Tarroni, G.: Anomaly detection through latent space restoration using vector quantized variational autoencoders. In: Proceedings of ISBI, pp. 1764–1767. IEEE (2021)
9. Masci, J., Meier, U., Cireşan, D., Schmidhuber, J.: Stacked convolutional autoencoders for hierarchical feature extraction. In: Honkela, T., Duch, W., Girolami, M., Kaski, S. (eds.) ICANN 2011. LNCS, vol. 6791, pp. 52–59. Springer, Heidelberg (2011). https://doi.org/10.1007/978-3-642-21735-7_7
10. Van den Oord, A., Vinyals, O., Kavukcuoglu, K.: Neural discrete representation learning. In: Proceedings of NIPS, pp. 6309–6318 (2017)

11. Peharz, R., et al.: Einsum networks: fast and scalable learning of tractable probabilistic circuits. In: Proceedings of ICML, pp. 7563–7574. PMLR (2020)
12. Peharz, R., et al.: Random sum-product networks: a simple and effective approach to probabilistic deep learning. In: Uncertainty in Artificial Intelligence, pp. 334–344. PMLR (2020)
13. Poon, H., Domingos, P.: Sum-product networks: a new deep architecture. In: Proceedings of ICCV Workshops, pp. 689–690. IEEE (2011)
14. Sato, D., et al.: A primitive study on unsupervised anomaly detection with an autoencoder in emergency head CT volumes. In: Medical Imaging 2018: Computer-Aided Diagnosis, vol. 10575, p. 105751P. International Society for Optics and Photonics (2018)
15. Schlegl, T., Seeböck, P., Waldstein, S.M., Langs, G., Schmidt-Erfurth, U.: f-AnoGAN: fast unsupervised anomaly detection with generative adversarial networks. Med. Image Anal. **54**, 30–44 (2019)
16. Seeböck, P., et al.: Exploiting epistemic uncertainty of anatomy segmentation for anomaly detection in retinal OCT. IEEE Trans. Med. Imaging **39**(1), 87–98 (2019)
17. Simarro Viana, J., de la Rosa, E., Vande Vyvere, T., Robben, D., Sima, D.M., et al.: Unsupervised 3D brain anomaly detection. In: Crimi, A., Bakas, S. (eds.) BrainLes 2020. LNCS, vol. 12658, pp. 133–142. Springer, Cham (2021). https://doi.org/10.1007/978-3-030-72084-1_13
18. Trapp, M., Peharz, R., Ge, H., Pernkopf, F., Ghahramani, Z.: Bayesian learning of sum-product networks. In: Proceedings of NeurIPS, pp. 6347–6358 (2019)
19. Wei, Q., Ren, Y., Hou, R., Shi, B., Lo, J.Y., Carin, L.: Anomaly detection for medical images based on a one-class classification. In: Medical Imaging 2018: Computer-Aided Diagnosis, vol. 10575, pp. 375–380. SPIE (2018)
20. Wild, C., Weiderpass, E., Stewart, B.W.: World Cancer Report: Cancer Research for Cancer Prevention. IARC Press (2020)
21. Zimmerer, D., Isensee, F., Petersen, J., Kohl, S., Maier-Hein, K.: Unsupervised anomaly localization using variational auto-encoders. In: Shen, D., et al. (eds.) MICCAI 2019. LNCS, vol. 11767, pp. 289–297. Springer, Cham (2019). https://doi.org/10.1007/978-3-030-32251-9_32

Image Translation Based Nuclei Segmentation for Immunohistochemistry Images

Roger Trullo[1(✉)], Quoc-Anh Bui[2], Qi Tang[3], and Reza Olfati-Saber[4]

[1] Sanofi, Chilly Mazarin, France
roger.trullo@sanofi.com
[2] Aix-Marseille University, Marseille, France
quoc-anh.bui@sanofi.com
[3] Sanofi, Bridgewater, NJ, USA
qi.tang@sanofi.com
[4] Sanofi, Cambridge, MA, USA
reza.olfati-saber@sanofi.com

Abstract. Numerous deep learning based methods have been developed for nuclei segmentation for H&E images and have achieved close to human performance. However, direct application of such methods to another modality of images, such as Immunohistochemistry (IHC) images, may not achieve satisfactory performance. Thus, we developed a Generative Adversarial Network (GAN) based approach to translate an IHC image to an H&E image while preserving nuclei location and morphology and then apply pre-trained nuclei segmentation models to the virtual H&E image. We demonstrated that the proposed methods work better than several baseline methods including direct application of state of the art nuclei segmentation methods such as Cellpose and HoVer-Net, trained on H&E and a generative method, DeepLIIF, using two public IHC image datasets.

Keywords: GAN · H&E · IHC · Nuclei · Segmentation

1 Introduction

H&E images are the most common modality of histopathology images since they can be stained quickly, economically and significance amount of microscopic anatomy is revealed. As a result, many nuclei segmentation methods have been developed recently leveraging pathologist's annotations and the advance in computer vision for H&E images [19]. Besides H&E, there is another popular type of image, the immunohistochemistry (IHC) image, which is commonly used to identify specific protein biomarkers and is complementary to H&E images [16]. IHC images play a central role in companion diagnostic tools for development of precision medicines. Several novel oncology therapies have been approved by regulatory agencies with a companion diagnostic device based on IHC images. On

A. Mukhopadhyay et al. (Eds.): DGM4MICCAI 2022, LNCS 13609, pp. 87–96, 2022.
https://doi.org/10.1007/978-3-031-18576-2_9

the other hand, the analysis of IHC images has been typically based on manual semi-quantitative methods, such as the 20X rule, for clinical decision making, which may suffer from subjectivity and inter and intra-rater variability. Several case studies suggest that deep learning based digital pathology algorithms may provide improved patient selection comparing to the current clinical standard [9]. At the core of these deep learning algorithms, there is a nuclei segmentation method, which relies on human expert annotations. To avoid the labor intensive manual annotation on IHC images and also to improve upon the performance of direct application of models trained on H&E images to IHC images, we proposed a two-step label-free approach where an IHC image is first translated into an H&E image utilizing unsupervised image-to-image translation methods, then for the second step, we apply existing methods that performed well on H&E images to virtually generated H&E images to obtain nuclei segmentation masks (Fig. 1).

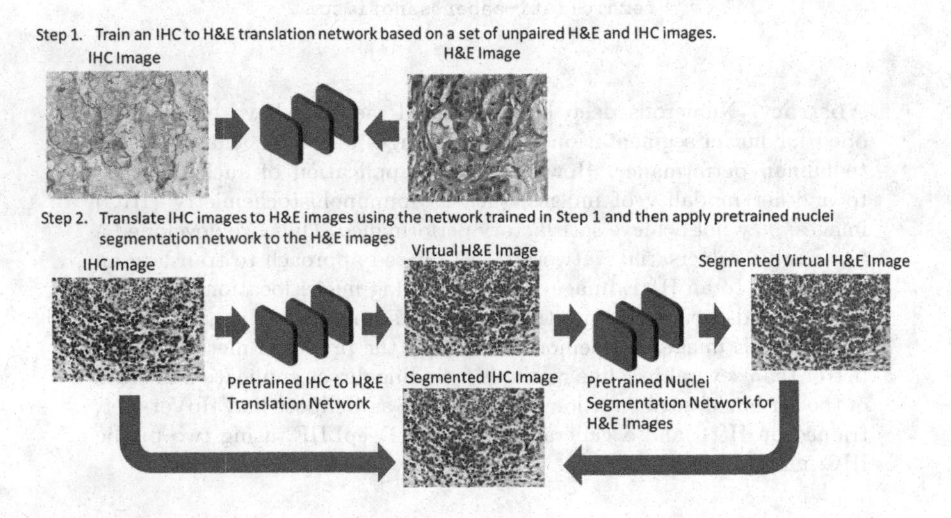

Fig. 1. Proposed IHC Segmentation Pipeline.

2 Related Work

Numerous cell morphology operation methods, such as color deconvolution, Otsu's thresholding, watershed, have been used either separately or in combination to segment nuclei in histology images [6]. With the advent of deep learning, it has been shown that supervised methods can outperform traditional cell morphology based methods [29] and traditional machine learning based methods [13]. However, a drawback of these methods and the supervised learning methods in general is that labor intensive manual annotation is required by trained human experts [1]. To avoid manual annotations, unsupervised deep learning methods have been developed [4,12] for nuclei segmentation, however these general methods were not tailored for IHC images. To develop nuclei segmentation models for IHC images with few annotations, researchers have leveraged the power of multiplex immunofluorescence (IF) staining to provide a ground truth of nuclei masks for the IHC nuclei segmentation through co-registration of IF images with IHC images [7] and then trained a conditional generative adversarial network (cGAN) [15] to generate nuclei masks.

3 Methodology

3.1 Translation from IHC Images to H&E Images

Image-to-image translation have achieved significant advancements since the advent of GAN. Application of GAN and its variations created a new field in digital pathology named virtual staining, where one or more image modalities can be virtually generated based on input of only one modality of image. Feasibility of virtual staining has been demonstrated across multiple histology image modalities [2,21,22,28,30] including H&E and IHC. When it comes to translation between H&E and IHC, almost all literature focused on the translation from H&E to IHC since H&E images were more commonly available for patients. However, for our purpose of leveraging well trained nuclei segmentation models for H&E images, we are interested in the reserve direction of the translation, that is, from IHC to H&E. Due to the challenges of registering IHC images with H&E images and high resolution of these types of images, generative methods that do not require paired images, such as CycleGAN [31] and U-GAT-IT [17], are methods of interest for the image translation task. These methods work by training simultaneously two GANs (two generator and two discriminator models). One generator translates IHC to H&E and other generator translates from H&E to IHC. In addition, in order to have pixel to pixel correspondence, a cycle consistency loss is introduced. This loss enforces the idea that when using as input an IHC image, the first generator will produce a virtual H&E image and if we use this virtual H&E image as input to the second generator, the produced image should match the input IHC image. On the other hand, U-GAT-IT extended the method by adding new normalization layers and by using attention modules which was claimed to improve the quality of the produced images. For both of these methods, we used the hyperparameters recommended by their authors including loss weights and learning rates. For CycleGAN we used a batch size of 10, whereas for U-GAT-IT we used a batch size of 2. This is due to the fact that U-GAT-IT uses more memory since it has fully connected layers. We trained CycleGAN for 30 epochs and U-GAT-IT for 20 epochs. The batch size was selected to use the full GPU memory available and the number of epochs was selected empirically by looking qualitatively at the generated virtual H&E tiles. We used an NVIDIA P100 GPU for training of the image translation models, which took around 4 h.

3.2 Nuclei Segmentation on Virtual H&E Images

Once virtual H&E images are generated, pretrained models that work well for H&E images can be applied to the virtual H&E images to obtain the nuclei segmentation masks. Since the image translation step kept the location and the morphology of each nuclei from the IHC images, the virtual H&E images perfectly matched with the IHC images at pixel level without the need of image registration. Thus, the nuclei segmentation masks obtained from the virtual H&E images can be directly used as the nuclei masks for the original IHC images. Without loss of generality, popular H&E image nuclei segmentation methods such as Cellpose [26], StarDist [24], and HoVer-Net [10], were considered for this step. Cellpose was trained on a diverse set of datasets including image sets BBBC038v1 and BBBC039v1 [3], image set MoNuSeg [18], and image set ISBI 2009 [5]. The mixed dataset consists of microscopy images, H&E images and fluorescence images, with about 1139 images for the task of nuclei segmentation. StarDist was trained on two H&E image datasets, MoNuSeg 2018 training dataset, which has 30 H&E images and around 22,000 nuclear boundary annotations from a diverse set of patients including breast cancer, kidney cancer, lung cancer, prostate cancer, bladder cancer, colon cancer and stomach cancer patients, and a TNBC dataset [20], which has 50 annotated H&E images from patients with triple negative breast cancer. Regarding HoVer-Net, it was trained on 41 H&E stained colorectal adenocarcinoma image tiles, containing 24,319 exhaustively annotated nuclei with associated class labels.

4 Experiments

We systematically evaluated both the image translation component and the nuclei segmentation component of the proposed approach to examine the impact of each component in comparison to two different types of baseline methods, the direct application of the nuclei segmentation models trained on H&E images to IHC images and an pre-trained nuclei segmentation model tailored for IHC images, DeepLIIF [7]. For all the experiments we used a P100 Nvidia GPU.

4.1 Datasets

The image translation component was trained based on an in-house IHC/H&E dataset. The nuclei segmentation component was tested and compared on two different IHC image datasets, the DeepLIIF testing dataset [7] and the LYON19 dataset [27].

In-house IHC/H&E Dataset. To train the generative model that transforms IHC images onto virtual H&E images we use an in-house dataset, that consists of 123 IHC whole slide images and 121 H&E whole slide images. The IHC images were stained to highlight cells expressing a protein target, which cannot be disclosed due to confidentiality reasons, however, it is shown in Fig. 2 that the generative model trained on this specific type of IHC dataset generalized well to other protein targets, such as CD3/CD8 and Ki67. For each of these slides, we randomly sample patches of size 256×256 and ended up with 2510 patches from IHC images and 2793 patches from H&E images.

DeepLIIF Testing Dataset. A public test set [7], which can be downloaded from https://zenodo.org/record/5553268, it includes 598 Ki67 IHC images of size 512×512 and 40x magnification from bladder carcinoma and non-small cell lung carcinoma slides. The expression of Ki67 is strongly associated with tumor cell proliferation and growth, and is widely used in routine pathological investigation as a proliferation marker.

LYON19 Dataset. Public testing set of LYON19 [27], which can be downloaded from https://zenodo.org/record/3385420, it contains 441 Regions of Interest (ROIs) from whole slide images (WSIs) of CD3/CD8 stained IHC images of lymphocytes from breast, colon, and prostate cancer patients' biopsy specimens. The 441 ROIs were selected from IHC images with 277 regular area ROIs, 59 clustered cell ROIs, and 105 artifact area ROIs. The ground truth of this dataset was not disclosed by the Grand Challenge competition. However, performance on this testing set can be obtained after submitting the predictions to the challenge here https://lyon19.grand-challenge.org/Submission/. The objective of the competition is to detect positive stained cells; in particular, the center of each positive cell. To accommodate for this, we computed our regular pipeline and then applied a simple thresholding procedure based on the HSI color space to classify cells as positively stained. We used the open source implementation provided by HistomicsTK [11]. Finally, we computed the centroid of each of those cells.

4.2 Baseline Methods and Evaluation Metrics

We compared the proposed methods with four baseline models: the pre-trained Cellpose model v2.0.5 [25], the pre-trained `2D_versatile_he` StarDist model v0.8.2 [23], the pre-trained HoVer-Net model v1.0 [14], and the pre-trained DeepLIIF model v1.1.2 [8].

The performance of all the methods will be evaluated based on Dice score, which measures pixel level segmentation performance. We also evaluated cell instance level performance metrics including accuracy, precision, recall and F1 score, conditional on a given Intersection over Union (IoU) threshold. Since the concept of true negatives in instance cell level detection is not valid, accuracy is computed from the number of true positives, TP, false positives, FP and false negatives, FN, as accuracy= TP/(TP + FP + FN) as it is commonly done in object detection.

4.3 Results

First, the proposed methods were compared against the baseline methods on the DeepLIIF testing dataset. CycleGAN plus Cellpose achieved the best results in terms of Dice score and precision at IoU= 0.5 as shown in Table 1. By translating IHC to H&E images using GAN based methods, such as CycleGAN and U-GAT-IT, the performance of Cellpose and HoVer-Net improved at least 0.1 in

Dice score comparing to directly application of these methods to IHC images. However, the combination of GAN based methods with StarDist failed to achieve improvement and even lead to worse performance comparing to StarDist alone. We also observed that CycleGAN performed better than or similarly to U-GAT-IT when combined with StarDist, Cellpose or HoVer-Net. By changing the thresholds for IoU, the accuracy curves in Fig. 3 confirm that the same conclusion holds against different thresholds.

Table 1. Performance of proposed methods against baseline methods on DeepLIIF testing dataset. Cell instance level segmentation accuracy, precision, recall and F1 score were evaluated at IoU = 0.5.

Method	Dice score	Accuracy	Precision	Recall	F1 score
DeepLIIF	0.66	0.20	0.31	0.37	0.34
StarDist	0.64	0.33	0.54	0.45	0.49
CycleGAN+StarDist	0.59	0.27	0.49	0.38	0.42
U-GAT-IT+StarDist	0.50	0.18	0.34	0.28	0.31
Cellpose	0.57	0.27	0.58	0.33	0.42
CycleGAN+Cellpose	**0.72**	**0.38**	0.63	**0.49**	**0.55**
U-GAT-IT+Cellpose	0.67	0.28	0.47	0.42	0.44
HoVer-Net	0.42	0.20	**0.64**	0.23	0.33
CycleGAN+HoVer-Net	0.68	0.34	0.62	0.42	0.44
U-GAT-IT+HoVer-Net	0.68	0.34	0.59	0.44	0.50

Figure 2 visualizes cell nuclei segmentation results of four selected tiles. From the top to the bottom are the input IHC, the virtually generated H&E based on CycleGAN, the ground truth, which was only available for the visualized tiles, and the cell masks of all the methods previously presented. Pixel-wise, the true positive (TP), false positive (FP) and false negative (FN) are represented by blue, red and green color respectively. The first row presents a testing Ki67 IHC image from the DeepLIIF testing set whereas the next 3 are the CD3/CD8 IHC images from the LYON19 dataset. For the Ki67 IHC image, the proposed methods perform similarly to DeepLIIF since DeepLIIF was trained specifically for this type of staining. However, for the CD3/CD8 IHC images, the proposed methods outperformed DeepLIIF and the direct application of Cellpose and HoVer-Net models pretrained on H&E images.

Next, we examined the performance of these methods on the LYON19 dataset. By submitting the model predictions to the Grand Challenge competition website, the F1 score were obtained and were tabulated in Table 2. Only CycleGAN was used for the image translation step since it has better performance than U-GAT-IT when tested in the DeepLIIF testing dataset. The proposed method performed similarly as the baseline methods except the combination of CycleGAN with HoVer-Net method, which lead to 4% higher F1 score

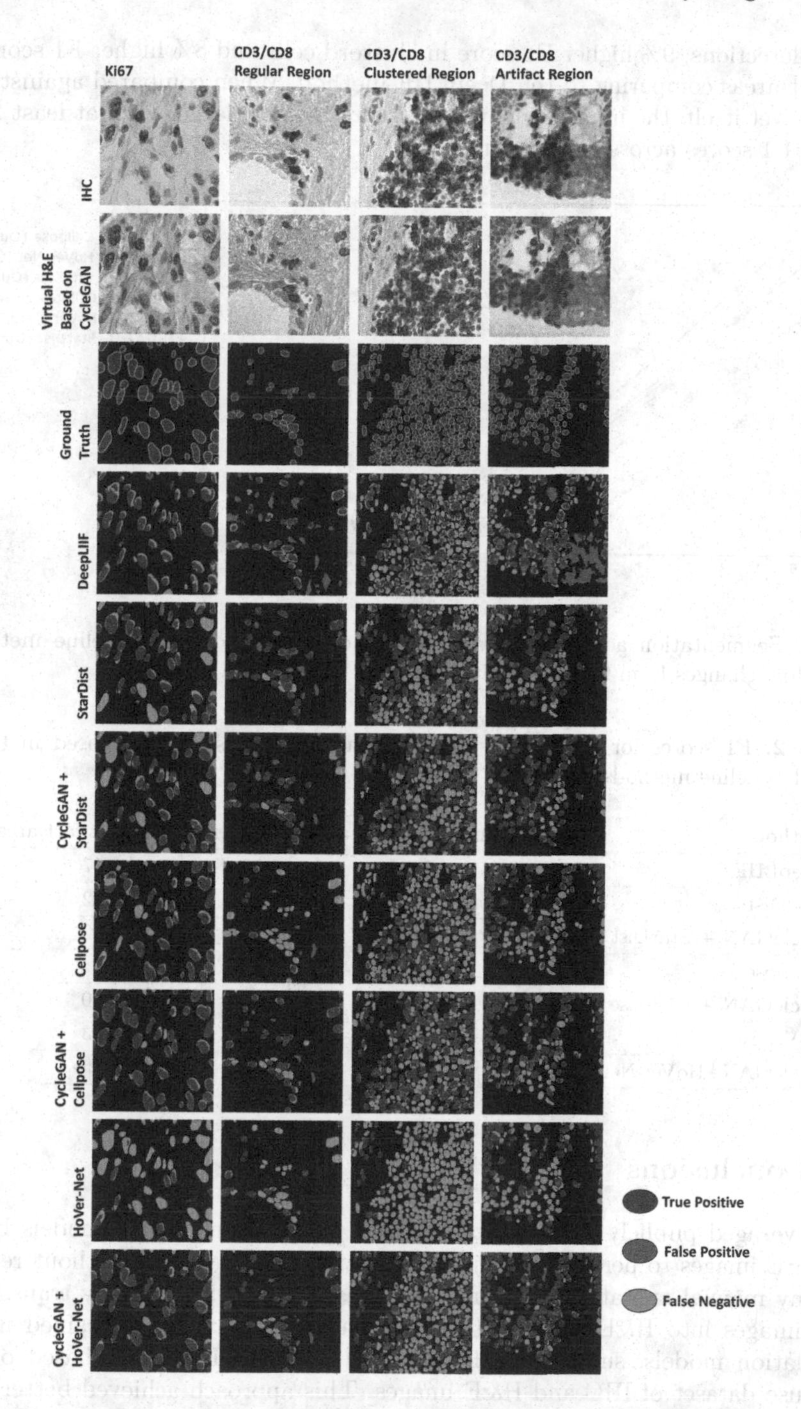

Fig. 2. Presentation of the proposed and the baseline methods' performance in Ki67 IHC images from DeepLIIF testing dataset and CD3/CD8 IHC images from LYON19 dataset.

in all detections, 9% higher F1 score in clusterd cells and 8% higher F1 score in artifact areas comparing to the DeepLIIF method. When compared against the HoVer-Net itself, the improvement in F1 score is much larger with at least 20% higher F1 scores across all categories.

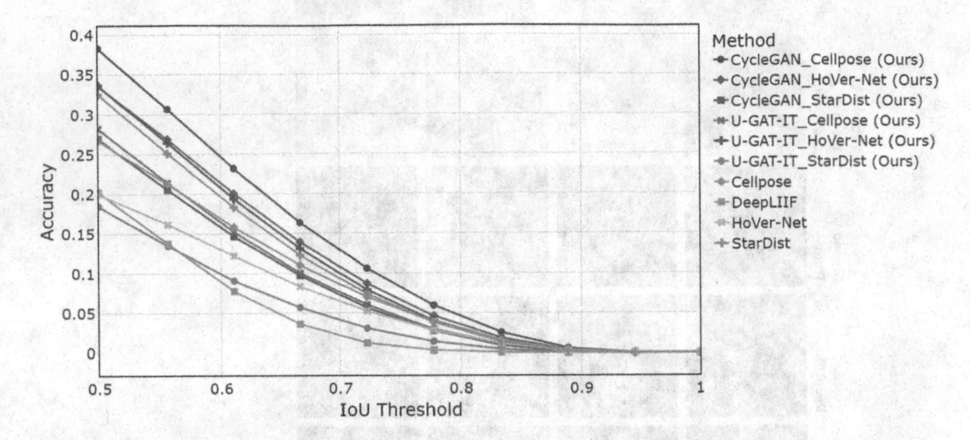

Fig. 3. Segmentation accuracy curves for proposed methods and baseline methods when IoU changes from 0.5 to 1 with a step size of 0.05.

Table 2. F1 scores for different types of segmentation tasks of proposed methods against baseline methods on LYON19 dataset.

Method	All detections	Regular areas	Clustered cells	Artifact areas
DeepLIIF	0.53	**0.64**	0.59	0.17
StarDist	0.54	0.59	0.63	0.22
CycleGAN + StarDist	0.54	0.59	0.63	0.22
Cellpose	0.54	0.53	**0.68**	0.24
CycleGAN + Cellpose	0.54	0.61	0.67	0.20
HoVer-Net	0.04	0.06	0.01	0.03
CycleGAN+HoVer-Net	**0.58**	0.63	**0.68**	**0.25**

5 Conclusions

We leveraged publicly available pre-trained nuclei segmentation models based on H&E images to perform nuclei segmentation in IHC images without requiring any manual anotation on them. This was achieved by virtually translating IHC images into H&E images. To enable such translation, we trained image translation models, such as CycleGAN and U-GAT-IT models, based on an in-house dataset of IHC and H&E images. This approach achieved better performance than several of the baseline methods, including direct application of pre-trained models based on H&E images, such as Cellpose and HoVer-Net, to IHC images, and a pre-trained model tailored for IHC image nuclei segmentation,

DeepLIIF, when tested on DeepLIIF testing dataset. Such improvement was less pronounced in the LYON19 dataset except for CycleGAN combined with HoVer-Net. An interesting finding is that HoVer-Net alone performed badly on LYON19 dataset however when combined with CycleGAN, it became the best performing method. These observations together with the findings that the combination of CycleGAN or U-GAT-IT with the StarDist method failed to achieve improved performance in both testing datasets suggest that if the pre-trained nuclei segmentation method for H&E images has strong generalizability to IHC images, translating IHC to H&E images may not lead to improved performance, such as the case for StarDist, however, on the other hand, if the pre-trained method utilized features specific to H&E images, image translation can lead to substantial improvement. Thus, adoption of the proposed method depends on the generalizability of the pre-trained models to the target image modality. Our work has been based completely on label-free IHC images. An interesting perspective is to combine our method with a few manual annotations in a semi-supervised learning setting which we believe can potentially improve the performance even further.

References

1. Abdolhoseini, M., Kluge, M.G., Walker, F.R., Johnson, S.J.: Segmentation of heavily clustered nuclei from histopathological images. Sci. Rep. **9**(1), 1–13 (2019)
2. Bayramoglu, N., Kaakinen, M., Eklund, L., Heikkila, J.: Towards virtual H&E staining of hyperspectral lung histology images using conditional generative adversarial networks. In: Proceedings of the IEEE International Conference on Computer Vision Workshops, pp. 64–71 (2017)
3. Caicedo, J.C., et al.: Nucleus segmentation across imaging experiments: the 2018 data science bowl. Nat. Methods **16**(12), 1247–1253 (2019)
4. Chen, M., Artières, T., Denoyer, L.: Unsupervised object segmentation by redrawing. Adv. Neural Inf. Process. Syst. **32** (2019)
5. Coelho, L.P., Shariff, A., Murphy, R.F.: Nuclear segmentation in microscope cell images: a hand-segmented dataset and comparison of algorithms. In: 2009 IEEE International Symposium on Biomedical Imaging: From Nano to Macro, pp. 518–521. IEEE (2009)
6. Di Cataldo, S., Ficarra, E., Acquaviva, A., Macii, E.: Automated segmentation of tissue images for computerized IHC analysis. Comput. Methods Programs Biomed. **100**(1), 1–15 (2010)
7. Ghahremani, P., et al.: Deep learning-inferred multiplex immunofluorescence for immunohistochemical image quantification. Nat. Mach. Intell. **4**(4), 401–412 (2022)
8. Ghahremani, P., et al.: Deepliif (2022). https://github.com/nadeemlab/DeepLIIF
9. Glass, B., et al.: 821 machine learning models can quantify cd8 positivity in lymphocytes in melanoma clinical trial samples. J. Immunother. Cancer **9**(Suppl 2), A859–A859 (2021)
10. Graham, S., et al.: Hover-net: simultaneous segmentation and classification of nuclei in multi-tissue histology images. Med. Image Anal. **58**, 101563 (2019)
11. Gutman, D.A., et al.: The digital slide archive: a software platform for management, integration, and analysis of histology for cancer research. Cancer Res. **77**(21), e75–e78 (2017). https://doi.org/10.1158/0008-5472.CAN-17-0629

12. Han, L., Yin, Z.: Unsupervised network learning for cell segmentation. In: de Bruijne, M., et al. (eds.) MICCAI 2021. LNCS, vol. 12901, pp. 282–292. Springer, Cham (2021). https://doi.org/10.1007/978-3-030-87193-2_27
13. Hatipoglu, N., Bilgin, G.: Cell segmentation in histopathological images with deep learning algorithms by utilizing spatial relationships. Med. Biol. Eng. Comput. **55**(10), 1829–1848 (2017). https://doi.org/10.1007/s11517-017-1630-1
14. Ho, M.Y., Chapman, V., Ali, Z., Graham, S., Vu, Q.D.: Hover-net (2022). https://github.com/vqdang/hover_net
15. Isola, P., Zhu, J.Y., Zhou, T., Efros, A.A.: Image-to-image translation with conditional adversarial networks. In: Proceedings of the IEEE conference on computer vision and pattern recognition, pp. 1125–1134 (2017)
16. Kaplan, K.: Quantifying IHC data from whole slide images is paving the way toward personalized medicine. MLO Med. Lab. Obs. **47**, 20–21 (2015)
17. Kim, J., Kim, M., Kang, H., Lee, K.H.: U-gat-it: unsupervised generative attentional networks with adaptive layer-instance normalization for image-to-image translation. In: International Conference on Learning Representations (2020). https://openreview.net/forum?id=BJlZ5ySKPH
18. Kumar, N., Verma, R., Sharma, S., Bhargava, S., Vahadane, A., Sethi, A.: A dataset and a technique for generalized nuclear segmentation for computational pathology. IEEE Trans. Med. Imaging **36**(7), 1550–1560 (2017)
19. Mahbod, A., et al.: CryoNuSeg: a dataset for nuclei instance segmentation of cryosectioned H&E-stained histological images. Comput. Biol. Med. **132**, 104349 (2021)
20. Naylor, P., Laé, M., Reyal, F., Walter, T.: Segmentation of nuclei in histopathology images by deep regression of the distance map. IEEE Trans. Med. Imaging **38**(2), 448–459 (2018)
21. Rivenson, Y., de Haan, K., Wallace, W.D., Ozcan, A.: Emerging advances to transform histopathology using virtual staining. BME Front. **2020** (2020)
22. Rivenson, Y., et al.: Virtual histological staining of unlabelled tissue-autofluorescence images via deep learning. Nat. Biomed. Eng. **3**(6), 466–477 (2019)
23. Schmidt, U., Weigert, M.: Stardist (2022). https://github.com/stardist/stardist
24. Schmidt, U., Weigert, M., Broaddus, C., Myers, G.: Cell detection with star-convex polygons. In: Medical Image Computing and Computer Assisted Intervention - MICCAI 2018–21st International Conference, Granada, Spain, 16–20 September 2018, Proceedings, Part II, pp. 265–273 (2018). https://doi.org/10.1007/978-3-030-00934-2_30
25. Stringer, C.: Cellpose (2022). https://github.com/MouseLand/cellpose
26. Stringer, C., Wang, T., Michaelos, M., Pachitariu, M.: Cellpose: a generalist algorithm for cellular segmentation. Nat. Methods **18**(1), 100–106 (2021)
27. Swiderska-Chadaj, Z., et al.: Learning to detect lymphocytes in immunohistochemistry with deep learning. Med. Image Anal. **58**, 101547 (2019)
28. Xu, Z., Moro, C.F., Bozóky, B., Zhang, Q.: Gan-based virtual re-staining: a promising solution for whole slide image analysis. arXiv preprint arXiv:1901.04059 (2019)
29. Yang, L., et al.: NuSeT: a deep learning tool for reliably separating and analyzing crowded cells. PLoS Comput. Biol. **16**(9), e1008193 (2020)
30. Zhang, Y., de Haan, K., Rivenson, Y., Li, J., Delis, A., Ozcan, A.: Digital synthesis of histological stains using micro-structured and multiplexed virtual staining of label-free tissue. Light Sci. Appl. **9**(1), 1–13 (2020)
31. Zhu, J.Y., Park, T., Isola, P., Efros, A.A.: Unpaired image-to-image translation using cycle-consistent adversarial networks. In: Proceedings of the IEEE International Conference on Computer Vision, pp. 2223–2232 (2017)

3D (c)GAN for Whole Body MR Synthesis

Daniel Mensing[1,2](✉), Jochen Hirsch[1], Markus Wenzel[1], and Matthias Günther[1,2,3]

[1] Fraunhofer MEVIS, Bremen, 28359 Bremen, Germany
daniel.mensing@mevis.fraunhofer.de
[2] mediri GmbH, 69115 Heidelberg, Germany
[3] Universität Bremen, 28359 Bremen, Germany

Abstract. Synthesis of images has recently seen many works that produce high-quality real world images. In the domain of medical imaging the application of deep generative models especially Generative Adversarial Networks (GANs) can be applied to many different tasks. Under the premise of the generation of high-quality images that match the distribution of the original data, the synthesized data can be used to increase the size of small datasets, or in combination with conditioning on meta data, to increase the size of underrepresented classes in the dataset. In this work we propose a model that generates 3D medical images. The model can easily be conditioned on meta data, for example available patient information. We evaluate the quality of the generated images and compare our model against the 3D-StyleGAN model which is also designed for 3D medical image synthesis.

Keywords: Generative adversarial networks · 3D Image Synthesis · Conditional GAN

1 Introduction

In this work we propose a GAN architecture for the generation of 3D volumetric images. The design decisions of the architecture were inspired by the findings of DCGAN [16] and FastGAN [14] which were then validated for 3D medical image synthesis through an ablation study. Additionally we propose to use linear conditioning in the generator and discriminator on available meta data. There is little work on 3D medical image synthesis, especially with high resolution greater than 64^3. This can partly be explained with the requirement of Graphical Processing Unit (GPU) memory imposed by the three dimensions of the data. Often this lack of GPU memory has to be compensated by reducing the number of feature maps or the depth of the network which makes the training more challenging. Some previous works tried to overcome this issue by synthesizing only a slab of the volume [5] or generating the slices of the volume separately [2]. Previous work that generate volumes directly by using 3D convolutions is often limited in

A. Mukhopadhyay et al. (Eds.): DGM4MICCAI 2022, LNCS 13609, pp. 97–105, 2022.
https://doi.org/10.1007/978-3-031-18576-2_10

size/resolution. The authors of 3D-StyleGAN build upon the well-known StyleGAN2 architecture and adapted it for three dimensions by significantly reducing the number of feature maps and the size of the latent vector, to generate T1 weighted MR images at 2 mm spatial resolution [7]. We investigated previously known best practices for GANs and evaluated their feasibility for 3D medical images through an ablation study. As a result this work proposes a GAN that generates synthetic whole body MR volumes with a size of 160 × 160 × 128. We achieve this by reducing each training batch to a single data sample which allows us to increase the number of feature maps. Additionally we show that the proposed model can easily be conditioned on meta data which further improves its performance. We compare our model, with and without conditioning, with the 3D-StyleGAN architecture.

2 Methods

Architecture and Training

We propose an architecture based on findings for effective GAN training and adapted them to and investigated them for the domain of 3D medical images. The overall architecture and many parts were introduced by FastGAN [14]. Figure 1 shows a simplified diagram of the architecture. One major motivation behind this decision is the low demand for training data by the FastGAN in combination with the low complexity of the network. All design choices were validated through an ablation study in which we investigated the influence of each part on the models performance. We used InstanceNorm [17] instead of batch normalization because the size of the data does not allow for large batch sizes thus rendering batch normalization less useful. The generator first maps the latent vector to the first feature maps which determine the size of the output through a transposed convolution layer. The main building block of the generator is depicted in Fig. 2 on the left. The remaining generator consists of five of these blocks, each of which doubles the resolution of the intermediate feature maps and a final convolutional layer which maps the feature maps to the number of output channels, in this case one channel for grayscale images.

The discriminator mirrors this architecture except that the resolution of the feature maps is reduced by a factor of two by using strided convolutions and that the activation function for each convolution is the Leaky ReLU function. Furthermore, there is no noise injection in the discriminator. The repeating building blocks of the discriminator are shown in Fig. 2 on the right. At the end of the discriminator the features are fed into a small convolutional network consisting of two layers which reduces the size further and serves as a critic whose output rates the input data as real or fake which then is used for the adversarial training.

Both, generator and discriminator employ Skip-Layer-Excitation layers, introduced in [14], which serve as a skip connection between two blocks at different depth of the network and helps to propagate the error to the first layers of the model. Another important part of the decoder is self-regularization due to decoders that decode the volume from the smallest feature map back to a volume with half the input size. This method was also introduced in [14] but we employ multiple decoders. We implemented one decoder, that decodes the feature maps to the whole volume, one that decodes to only one part of the input volume and one that only decodes the abdomen section to ensure high detail in this region. The decoder networks use transposed convolutions for the upsampling and no conditioning on meta data regardless of the conditioning in the generator and discriminator. The loss for the generator is the output of the critic, while the loss for the discriminator is the sum of the adversarial hinge loss [13] and separate reconstruction losses for each decoder, which were the mean absolute error between the decoded image and the interpolated or cropped part of the real image. Other methods used during the training process were exponential moving average of the generator weights [19], early stopping and learning rate decay.

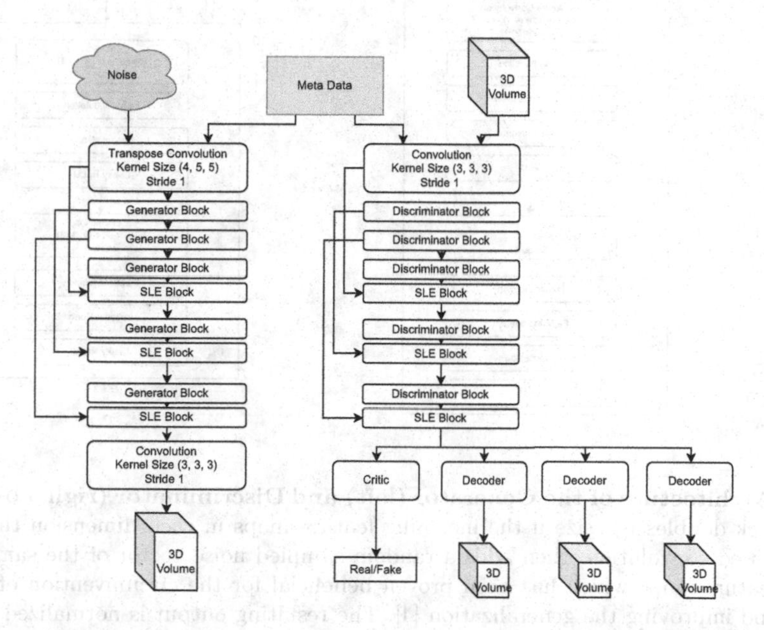

Fig. 1. Architecture of the Generator (left) and Discriminator (right) networks. Simplified view of the architecture of our proposed GAN. The meta data input for both models is optional. A detailed view of the generator and discriminator blocks is provided in Fig. 2. In general the architecture is inspired by the overall architecture proposed in [14]. Skip Layer Excitation (SLE) blocks are used to propagate the error to the first layers of the model.

Conditioning on Meta data

Many use cases benefit from the ability to generate data conditioned on given attributes. The following patient information were used for the conditional 3D image synthesis: *age*, *sex*, *weight* and *height.* For conditioning, we added a Feature-wise Linear Modulation (FiLM) layer [15] between each convolutional layer and the noise injection layer. This layer affine transforms the intermediate feature maps with two learned parameter vectors γ and β, which are provided by an encoder, which is trained together with the model, that is shared through all FiLM layers in the model (generator and discriminator each have their own). For this experiment, we binary-encoded the meta data and concatenated all binary vectors which then serves as input for the encoder. If the network shall be conditioned on additional input data, a FiLM modulation layer follows between the convolution layer and the noise injection layer. The linear conditioning with meta data was shown to be beneficial for image segmentation by [12].

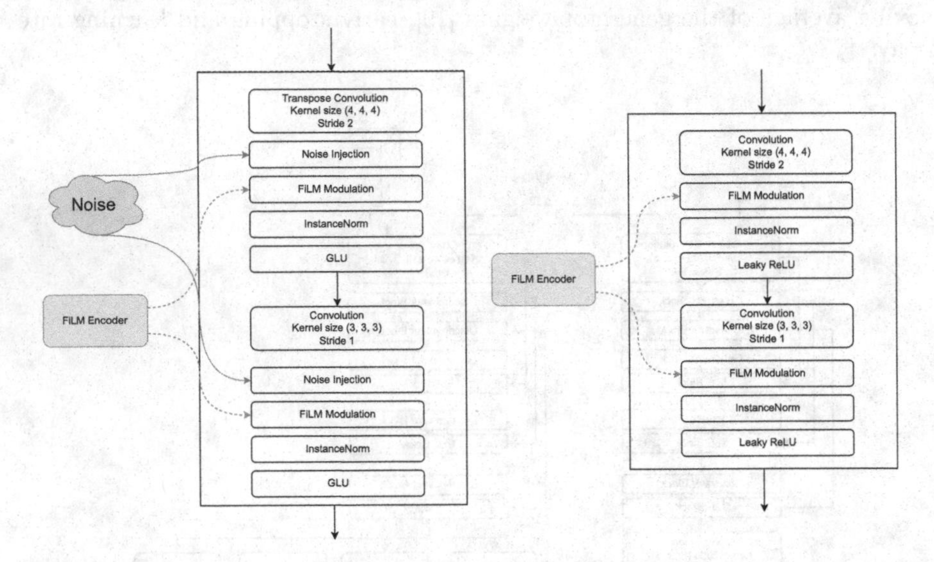

Fig. 2. Architecture of the Generator (left) and Discriminator (right) blocks. Each block doubles the size if the incoming feature maps in each dimension through a transposed convolution, then adds a random sampled noise vector of the same size as the feature maps which has been proven beneficial for the circumvention of overfitting and improving the generalization [4]. The resulting output is normalized by an Instance Normalization [17] layer and a Gated Linear Unit (GLU) [3] operation serves as activation function. The same structure is repeated once more with the transposed convolution replaced by a regular convolution

Data

The used dataset consists of 10828 whole body MR volumes obtained as part of the MR Imaging Study within the German National Cohort Study (GNC, 2014-2019) [1] from volunteers. The data was acquired on MAGNETOM Skyra 3T (Siemens Healthineers, syngo VD13C) systems with a two-point Dixon volumetric interpolated breath-hold examination (VIBE) T1 weighted sequence. We used the so-called "opposed phase" contrast (TE = 1.23 ms). The volumes were acquired by axial acquisition with in-plane matrix 320 × 260 (resolution 1.4 × 1.4 mm^2) and a slice thickness of 3 mm. The volume consists of four acquired table positions with a total of 316 slices which were then resampled and cropped to 160 × 160 × 128 which doubles the voxel size but therefore reduces the size of each volume roughly by half in order to fit the volume on the GPU. All intensity values were scaled to the range of [−1, 1]. We used half of the dataset for training and the other half for evaluation.

Evaluation

For the evaluation of the quality of the generated volumes we used the slice-wise Fréchet Inception Metric proposed in [7]. Since the Fréchet Inception Distance (FID) [6] is calculated from features extracted from a Inception V3 network pretrained on the Imagenet dataset, which is a 2D dataset, we calculate the FID score for the center slice for each orientation. Additionally, the Multi-Scale Structured Similarity Measure (MS-SSIM) and the Maximum Mean Discrepancy (MMD) were used for the evaluation. The MMD measures the distance between two distributions and was calculated batch-wise as proposed in [11,18] and [7]. The MS-SSIM measures the structural similarity between two samples at different scales and can be used to evaluate the diversity of the generated images [18].

3 Results

The results shown in Table 1 show that our model without conditioning has a much lower MMD and FID and higher MS-SSIM than the trained 3D-StyleGAN. A comparison between a sample generated by the 3D-StyleGAN and our unconditional model is shown in Fig. 3. In comparison our model with conditioning on meta data results in even slightly better scores across almost all metrics. The 3D-StyleGAN was trained with 1 mm-fd16 configuration which was the only one that allows to generate volumes at the size of 160 × 160 × 128. The only change to the configuration was the output size of the base layer which was adapted from 5 × 6 × 7 to 5 × 5 × 4 to result in the desired output size.

Table 1. Results. The table shows the MMD, the MS-SSIM between whole volumes of generated and real data. The FID was calculated for the center slice of the volume in Axial (FID Ax.), Sagittal (FID Sag.) and Coronal (FID Cor.) orientation between generated and real samples. ↓ means that a lower metric score is better and ↑ shows that a higher value is better.

	MMD ↓	MS-SSIM ↑	FID (Ax.) ↓	FID (Sag.) ↓	FID (Cor.) ↓
3D-StyleGAN	47307 ± 13162	0.162 ± 0.004	362.5 ± 1.6	373.7 ± 15.9	431.6 ± 11
Ours	12086 ± 641	0.409 ± 0.004	**71.2** ± 1.0	43.3 ± 5.7	106.4 ± 23.7
Ours Conditional	**10589** ± 333	**0.439** ± 0.001	76.5 ± 2.5	**38.4** ± 10.2	**81.6** ± 22.5

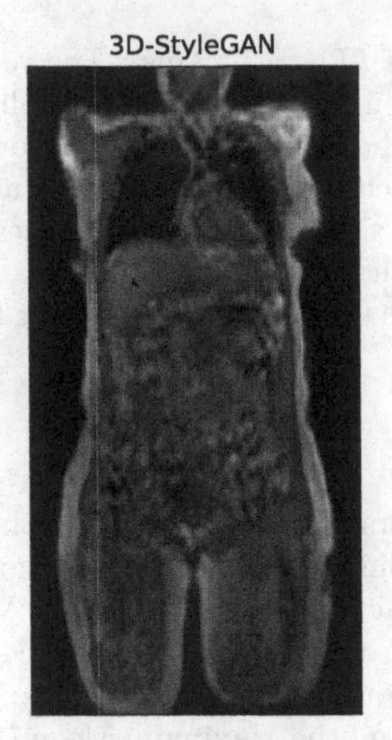

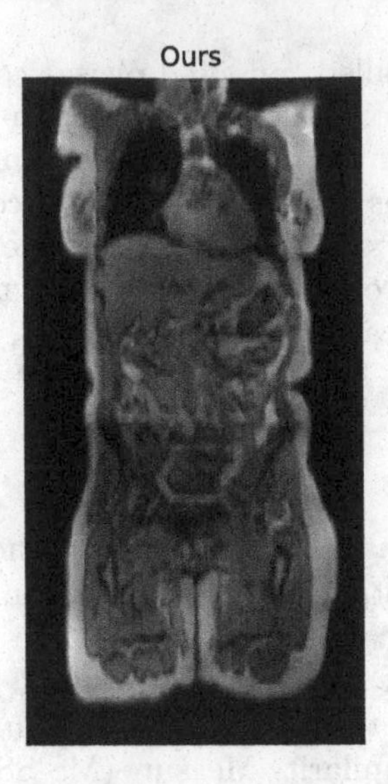

Fig. 3. Comparison between 3D-StyleGAN and ours. This figure shows the center slice in coronal orientation of two samples, on the left side generated by 3D-StyleGAN and on the right side by our proposed model. Both samples were generated unconditionally.

Conditional generation of 3D images

The results of the conditioning process were not evaluated separately. A visual inspection of the conditionally generated volumes showed that these were consistent with the meta data they were conditioned on which can be seen in Fig. 4. Depicted are generated volumes from the same latent vector and with different meta data conditioning. The images show the center slice of male and female volumes with different weights from 60 to 110 Kg. The other two remaining attributes stayed fixed.

4 Discussion

We propose a GAN architecture for 3D medical image synthesis that uses best practices for GAN training known from other domains. In order to leverage the often limited datasets available for medical imaging we added self regularization by adding decoders to the discriminator as proposed and justified in [14]. We assess our models performance with commonly used metrics for the evaluation of GANs and compare these against the 3D-StyleGAN architecture at the same resolution. In Table 1 we show that our model outperforms the 3D-StyleGAN in every metric. A possible explanation of the in general low MS-SSIM score across all compared models may be partially explained by the fact, that the training data has not been registered and therefore exhibits variation in size and scale of the samples. Since the MS-SSIM compares the structural similarity of spatially

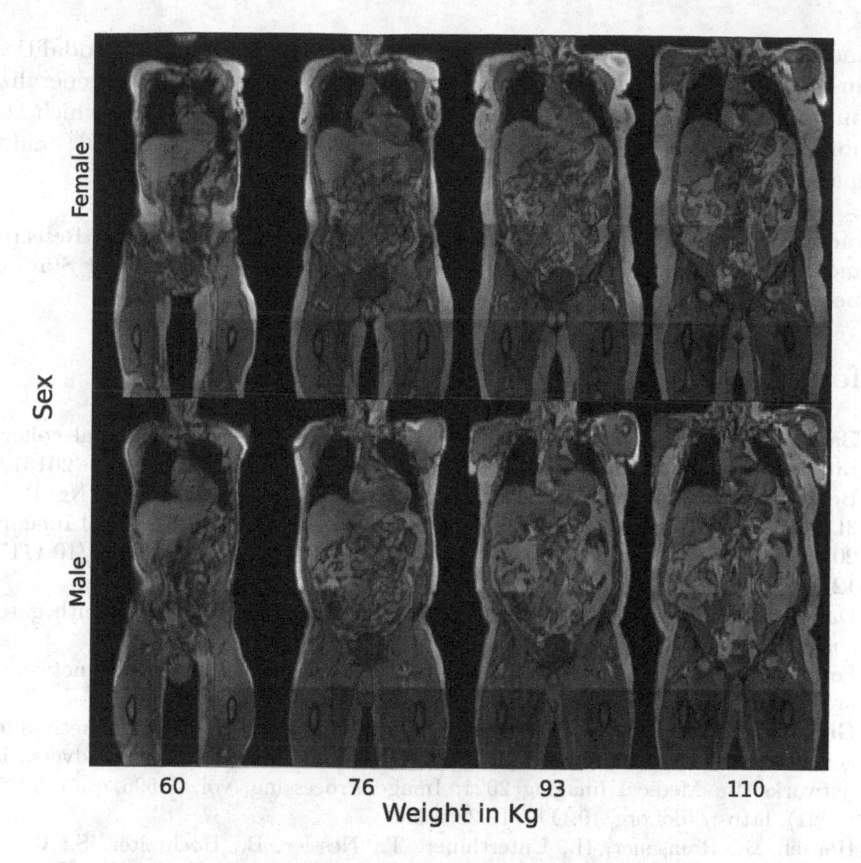

Fig. 4. Conditionally generated volumes. Center slices of volumes generated from the same latent vector with different conditional inputs. The upper row shows samples for the attribute *sex* set to female and the lower row with the attribute set to male. Both rows show the variation for the *weight* attribute ranging from 60 to 110 Kg. The remaining attributes *height* and *age* were set to 170 cm and 40 years old respectively. The stitching artefact that can be seen was caused by movements between the four acquisitions of which the volume is put together and was also learned by the GAN.

close voxels and the corresponding voxel in two compared samples can be at spatially different locations due to the patients size, this is a possible explanation for low MS-SSIM scores. [7] argued their model was not able to generate realistic images at 1 mm isotropic resolution which translates to an image size of $160 \times 192 \times 224$. Since the size of our images ($160 \times 160 \times 128$) is in between the size of their successful experiments and their failure case, we can only deduct that our model results in lower metric scores at the reported size. The conditioning on patient information has to be investigated further in regard of the independence of the different attributes and if the conditionally generated 3D images are plausible for the used meta data. Further improvements to the proposed architecture could stem from the StyleGAN [8–10] models which propose differentiable data augmentation or weight modulation. Further experiments with other modalities, organs and image sizes are needed to show the ability of the model to generalize beyond the trained data. Very recently HA-GAN was published in which the authors synthesize chest CT and brain MR images with a size of 256^3 and a comparison is left for future work.

Acknowledgement. We received grant money from the U Bremen Research Alliance/AI Center for Health Care, financially supported by the Federal State of Bremen.

References

1. Bamberg, F., et al.: Whole-body MR imaging in the German national cohort: rationale, design, and technical background. Radiology **277**(1), 206–220 (2015)
2. Bergen, R.V., Rajotte, J.F., Yousefirizi, F., Klyuzhin, I.S., Rahmim, A., Ng, R.T.: 3D PET image generation with tumour masks using TGAN. In: Medical Imaging 2022: Image Processing, vol. 12032, p. 120321P (2022). https://doi.org/10.1117/12.2611292
3. Dauphin, Y.N., Fan, A., Auli, M., Grangier, D.: Language modeling with gated convolutional networks. arXiv (2016)
4. Feng, R., Zhao, D., Zha, Z.: On noise injection in generative adversarial networks. arXiv (2020)
5. Granstedt, J.L., Kelkar, V.A., Zhou, W., Anastasio, M.A.: SlabGAN: a method for generating efficient 3D anisotropic medical volumes using generative adversarial networks. In: Medical Imaging 2021: Image Processing, vol. 11596, p. 1159617 (2021). https://doi.org/10.1117/12.2581380
6. Heusel, M., Ramsauer, H., Unterthiner, T., Nessler, B., Hochreiter, S.: GANs trained by a two time-scale update rule converge to a local Nash equilibrium. arXiv (2017)
7. Hong, S., et al.: 3D-StyleGAN: a style-based generative adversarial network for generative modeling of three-dimensional medical images. arXiv (2021)
8. Karras, T., Aittala, M., Hellsten, J., Laine, S., Lehtinen, J., Aila, T.: Training generative adversarial networks with limited data. arXiv (2020)
9. Karras, T., Laine, S., Aila, T.: A style-based generator architecture for generative adversarial networks. arXiv (2018)
10. Karras, T., Laine, S., Aittala, M., Hellsten, J., Lehtinen, J., Aila, T.: Analyzing and improving the image quality of StyleGAN. arXiv (2019)
11. Kwon, G., Han, C., Kim, D.S.: Generation of 3D brain MRI using auto-encoding generative adversarial networks. arXiv (2019)
12. Lemay, A., Gros, C., Vincent, O., Liu, Y., Cohen, J.P., Cohen-Adad, J.: Benefits of linear conditioning with metadata for image segmentation. arXiv (2021)
13. Lim, J.H., Ye, J.C.: Geometric GAN (2017). https://doi.org/10.48550/ARXIV.1705.02894, https://arxiv.org/abs/1705.02894
14. Liu, B., Zhu, Y., Song, K., Elgammal, A.: Towards faster and stabilized GAN training for high-fidelity few-shot image synthesis. arXiv (2021)
15. Perez, E., Strub, F., Vries, H.D., Dumoulin, V., Courville, A.: FiLM: visual reasoning with a general conditioning layer. arXiv (2017)
16. Radford, A., Metz, L., Chintala, S.: Unsupervised representation learning with deep convolutional generative adversarial networks. arXiv (2015)

17. Ulyanov, D., Vedaldi, A., Lempitsky, V.S.: Instance normalization: The missing ingredient for fast stylization. CoRR abs/1607.08022 (2016), http://arxiv.org/abs/1607.08022
18. Volokitin, A., et al.: Modelling the distribution of 3D brain MRI using a 2D slice VAE. arXiv (2020)
19. Yazıcı, Y., Foo, C.S., Winkler, S., Yap, K.H., Piliouras, G., Chandrasekhar, V.: The unusual effectiveness of averaging in GAN training. arXiv (2018)

Airway Measurement by Refinement of Synthetic Images Improves Mortality Prediction in Idiopathic Pulmonary Fibrosis

Ashkan Pakzad[1(✉)], Mou-Cheng Xu[1], Wing Keung Cheung[1], Marie Vermant[2,3], Tinne Goos[2,3], Laurens J. De Sadeleer[2,3], Stijn E. Verleden[2,4], Wim A. Wuyts[2,3], John R. Hurst[5], and Joseph Jacob[1,5]

[1] Centre for Medical Image Computing, University College London, London, UK
a.pakzad@cs.ucl.ac.uk

[2] BREATHE, Department of Chronic Diseases and Metabolism, KU Leuven, Leuven, Belgium

[3] Department of Respiratory Diseases, Unit for interstitial lung diseases, University Hospitals Leuven, Leuven, Belgium

[4] Antwerp Surgical Training, Anatomy and Research Centre (ASTARC), Faculty of Medicine and Health Sciences, University of Antwerp, Antwerp, Belgium

[5] UCL Respiratory, University College London, London, UK
https://ashkanpakzad.github.io

Abstract. Several chronic lung diseases, like idiopathic pulmonary fibrosis (IPF) are characterised by abnormal dilatation of the airways. Quantification of airway features on computed tomography (CT) can help characterise disease severity and progression. Physics based airway measurement algorithms that have been developed have met with limited success, in part due to the sheer diversity of airway morphology seen in clinical practice. Supervised learning methods are not feasible due to the high cost of obtaining precise airway annotations. We propose synthesising airways by style transfer using perceptual losses to train our model: Airway Transfer Network (ATN). We compare our ATN model with a state-of-the-art GAN-based network (simGAN) using a) qualitative assessment; b) assessment of the ability of ATN and simGAN based CT airway metrics to predict mortality in a population of 113 patients with IPF. ATN was shown to be quicker and easier to train than simGAN. ATN-based airway measurements showed consistently stronger associations with mortality than simGAN-derived airway metrics on IPF CTs. Airway synthesis by a transformation network that refines synthetic data using perceptual losses is a realistic alternative to GAN-based methods for clinical CT analyses of idiopathic pulmonary fibrosis. Our source code can be found at https://github.com/ashkanpakzad/ATN that is compatible with the existing open-source airway analysis framework, AirQuant.

Supplementary Information The online version contains supplementary material available at https://doi.org/10.1007/978-3-031-18576-2_11.

A. Mukhopadhyay et al. (Eds.): DGM4MICCAI 2022, LNCS 13609, pp. 106–116, 2022.
https://doi.org/10.1007/978-3-031-18576-2_11

Keywords: Generative model evaluation · Style transfer · Computed tomography · Airway measurement · Bronchiectasis · Idiopathic pulmonary fibrosis

1 Introduction

Chronic lung disease is one of the leading causes of morbidity and mortality across the world. As smoking rates in the developing world increase, the prevalence of chronic lung disease is set to rise. Interstitial lung diseases (ILD) are characterised by inflammation and scarring of the lung and the incidence of ILD continues to increase [25].

A subset of ILDs are characterised by lung fibrosis, with idiopathic pulmonary fibrosis (IPF) having the worst prognosis of all the fibrosing ILDs [4]. In IPF the airways are pulled open by fibrotic contraction of the surrounding connective tissue. Computed tomography (CT) imaging is used to visualise airway structure. In IPF the presence of dilated airways in the lung periphery on CT, termed traction bronchiectasis, is a disease hallmark.

When assessing disease severity in IPF, physiologic measurements are typically used. However these are associated with a degree of measurement variability. It has been postulated that combining imaging measures of airway abnormality with lung function measurements could help improve estimation of disease severity in IPF [18]. Importantly, better measures of disease severity would benefit cohort enrichment of subjects into therapeutic trials.

Lung damage in IPF progresses from the distal lung towards the centre of the lung [15]. As a result, the earliest signs of lung damage are seen in the smaller airways. Yet these airways are typically the most challenging to quantify. Airway measurement is complicated by partial volume effects that result in smaller airways having a blurred contour to their walls. Measurement challenges are compounded by variations in CT image acquisition including different reconstruction kernels, scan parameters and scanner models as well as the underlying pathology affecting the lung.

Physics based airway measurement algorithms tend to perform sub optimally when measuring the lumens of small airways [3,12]. Identifying airway walls can also be challenging. Airway paths often run in tandem with those of the pulmonary artery. Consequently, in regions when the pulmonary artery abuts the airway wall, identification of the contour of the outer airway wall is compromised.

1.1 Related Work

Deep learning frameworks have been applied to the measurement of airways in the lung in a bid to improve measurement accuracy. However, these machine learning methods are extremely data hungry and can be biased towards the training data sample [10]. Synthetic data by way of generative models has been employed to improve the training of deep learning models. This helps overcome the data limitations that are ubiquitous to medical imaging studies [24].

A state of the art method in measuring airway lumen radius and wall thickness on CT imaging, simGAN [16,21], takes labelled simplistic representations of airway patches (synthetic images) and aims to transforms them in to the emulations of real airways by generative adversarial training (GAN) [6]. These are then used for supervised training of a convolutional neural regressor (CNR) which learns to measure airway radius and wall thickness and ultimately to run inference on real CT images.

The driving loss for realism in simGAN is cross-entropy loss computed on the classifications of the discriminator. For successful synthetic refinement by image transformation, the synthetic and refined images must have good correspondence in their shared label. To this end, a per-pixel $||l||_1$ regularisation loss is applied between input and output of the refiner.

GAN training is inherently unstable with mode collapse complicating and lengthening training times. As an alternative strategy, in this paper we propose the first use of perceptual losses to generate labelled synthetic airway images. Perceptual loss functions have been applied to image style transfer and super-resolution tasks [11]. We explore the clinical benefits of learning from perceptual loss generated synthetic data in mortality prediction.

2 Methods

In the first part of our study we generate synthetic airway patches that demonstrate realistic airway characteristics. In tandem, we segment the airways on clinical CT scans of a cohort of IPF patients. We train our Airway Transfer Network (ATN) to transform our synthetic images to refined images across our synthetic and real datasets by optimising for perceptual losses. We then compare the results of ATN with simGAN. A CNR is trained on the resultant refined datasets for the purpose of inference on real CT airways. We compare the two refiner models qualitatively. We compare ATN and simGAN against the full width at half maximum edgecued segmentation limited (FWHMesl) technique as implemented in [20], originally by [12]. The FWHMesl technique is widely used in the literature as the reference for comparison of previous airway measurement methods [7,16,26]. In our clinical comparison, we examine which of the three methods of airway measurement provides the best and most consistent association with mortality on CT scans of patients with IPF.

Airway segmentation was performed using a 2D dilated U-Net [27] trained on CT scans in 25 IPF and healthy individuals [17]. We extract orthogonal airway patches for all segmented airways. We parameterise airway labels as two ellipses that share centre and rotation, resulting in 7 parameters for each patch: inner airway wall major and minor axis radii R_A and R_B; outer airway wall major and minor axis radii W_A and W_B; centre coordinates C_x and C_y; and rotation θ. Due to the phase in θ, for the purposes of CNR training the rotation angle is converted into a double angle representation [13].

Once the refiner model has been trained, its output is used to train a CNR by supervised learning to regress to target airway labels. The inner and outer airway wall measures are then derived. All deep learning methods were implemented in pytorch [19] and CT image processing was done using the open source airway analysis framework known as AirQuant [17]. We release our code open source[1].

2.1 Airway Synthesis

Details of airway parameters and synthesis pipeline have been previously described [16]. Airway characteristics are sampled from a set of distribution parameters informed by [23]. We deviate from these parameters in two ways. First, we use an airway lumen radius (LR) interval of [0.3, 6] to permit measurement of smaller airways. Second, we use an airway wall thickness $[0.1 \cdot LR + 0.2, 0.3 \cdot LR + 0.8]$ mm to reflect the lack of airway wall thickening in IPF. We add four further parameters: (i) parameters for the airway centre determined by a normal distribution $X \sim N(0, 1)$ mm to account for airway skeletons that are not perfectly positioned within the centre of the airway lumen. (ii) $p = 0.4$ that an adjacent airway of similar diameter is randomly added. This is performed to accommodate airway patches close to airway bifurcations and to train the CNR to correctly identify the airway in the centre of the patch. (iii) We model our airways as ellipsoids, we achieve this by an ellipsoidness characteristic, sampled from a uniform distribution, $X \sim U(0.9, 1)$ which determines the ratio in major and minor radii of the ellipse. (iv) Uniformly random rotation applied to the airway in the horizontal axis. We include our synthetic airway generator and configuration parameters in the open-source code repository.

2.2 Perceptual Losses

We implement perceptual losses for computing high level perceptual differences between synthetic and real images as described by [11]. These losses are computed by comparing the activations in particular layers, j of a pretrained convolutional neural network (CNN), ϕ between a pair of images. Different activation layers of a trained CNN learn to represent different image features on the same sampled patch. In minimising for perceptual losses we are looking to reduce the differences in the activation of these layers between the refiner output and some objective image. For each calculation of perceptual losses on a synthetic input image, x we have a refiner prediction, $\hat{y}$. As a modification of the original style transfer implementation [11], a randomly chosen real image is selected as the style target, y_s. Perceptual losses are then calculated and summed for different layers ϕ_j.

We utilise feature reconstruction loss. This is defined as the mean euclidean distance between activations of the input and output images of the refiner, where C, H, and W are the number of channels, height and width of layer j respectively. We use a VGG-16 [22] network pretrained on the ImageNet dataset [2] in our calculations of style and feature losses.

$$l_{feat}^{\phi,j}(\hat{y}, x) = \frac{1}{C_j H_j W_j} \|\phi_j(\hat{y}) - \phi_j(x)\|_1 \quad (1)$$

[1] https://github.com/ashkanpakzad/ATN.

We also employ style reconstruction loss, which considers those features that tend to be activated together between the refiner output and the given style target image, a random real airway, where G_j^ϕ is the gram matrix for a given layer j of ϕ as described in [5].

$$l_{style}^{\phi,j}(\hat{y}, y_s) = \frac{1}{C_j H_j W_j} \|G_j^\phi(\hat{y}) - G_j^\phi(y_s)\|_1 \tag{2}$$

2.3 Clinical Data

We examined CT images from 113 IPF patients diagnosed at the University Hospitals Leuven, Belgium. CTs were evaluated by an experienced chest radiologist (author JJ) for quality i.e. absence of breathing artefacts and infection. The quality of the automated segmentation was also visually inspected to ensure contiguous airway segmentations without oversegmentation blowouts. Airway segmentations were also required to reach the sixth airway generation in the upper and lower lobes to be selected for analysis. Pulmonary function tests were considered if they occurred within 90 days of the CT scan: Forced Vital Capacity (FVC, n = 111)); diffusing capacity of the lung for carbon monoxide (DLco, n=103).

The trachea and first generation bronchi were excluded from analysis. We define an airway segment as the length of airway that runs between airway branching points or an airway endpoint. All airway segments were pruned by 1 mm at either end to avoid bifurcating patches. 80×80 pixel size orthogonal airway patches were linearly interpolated with a pixel size of 0.5×0.5 mm from the CT at 0.5 mm intervals along each segment. This resulted in a final set of 546,790 real CT-derived airway patches. A synthetic dataset of 375,000 patches was generated to train our refiner.

27% of patients were female. 74% of patients had smoked previously. The median patient age was 71, with 57% of patients having died. All patients had received antifibrotic drug treatment.

Measures of intertapering, intratapering [14] and absolute airway volume were derived from the airway measurements for each airway segment. **Segmental intertapering** represents the relative difference in diameter of an airway segment when compared to its parent segment. Segmental intertapering is calculated as the difference in mean diameter, $\bar{d}$ of an airway segment and its parent segment, $\bar{d_p}$, divided by the mean diameter of the parent segment. **Segmental intratapering** is the gradient of change in diameter of the airway segment relative to the diameter of the origin of the segment[2]. Segmental intratapering is computed by dividing the gradient, m by the zero-intercept, c of a line $y = mx+c$ fitted to the diameter measurements of an airway segment. **Segmental volume** is computed by summing area measurements along an airway segment, and multiplying this value by the measurement interval, i.e. an integration of area along the segment's length.

[2] Segments are considered to be oriented from the centre of the lung to the periphery. Accordingly, measurement of the airway origin beings at the end closest to the trachea.

$$intertapering = \frac{\bar{d_p} - \bar{d}}{\bar{d_p}} \quad (3)$$

$$intratapering = \frac{-m}{c} \quad (4)$$

Univariable and multivariable Cox proportional hazards models were used to examine patient survival. Multivariable models included patient age (years), gender, smoking status (never vs ever) and either FVC or DLco (as measures of disease severity) as covariates. The goodness of fit of the model was denoted by the concordance index [8]. A p-value of <0.05 was considered statistically significant.

2.4 Implementation Details

We use the same refiner architecture as in [16,21], the refiner is a purely convolutional network with four repeating 3×3, 64 feature ResNet blocks [9]. The measurement CNR, described in [16], is a convolutional network that feeds into two fully connected layers to learn the airway ellipse parameters. Instead of the custom CNR loss described in [16], we implemented a mean square error (MSE) loss for regressing to the airway ellipse parameters.

Synthetic images were generated to 0.5×0.5 mm pixel size making 80×80 pixel patches, corresponding to the real patch generation noted in Sect. 2.3. All images were standardised and augmented on the fly, adding random Gaussian noise [25, 25] Hounsfield units, random levels of Gaussian blurring with standard deviation scalled in the interval [0.5, 0.875] and random flipping ($p = 0.2$). We apply random scaling on real images only, in the interval [0.75, 1.25] to increase diversity in airway size. Finally, a centre crop was applied to make a 32×32 pixel input patch.

Both simGAN and ATN models were trained for 10000 steps, where the simGAN refiner had 50 training iterations and the discriminator 1 iteration for every 1 step. The simGAN discriminator was implemented as described in the original method, with a memory buffer and local patch discrimination [21]. We used Weights & Biases for experiment tracking [1].

Figure demonstrates the overall method employed here as well as the ATN and CNR architecture.

3 Results

We implemented all training on an NVIDIA GeForce RTX 2070 graphical processing unit with a batch size of 256, learning rate of 0.001, $||l||_1$ regularisation

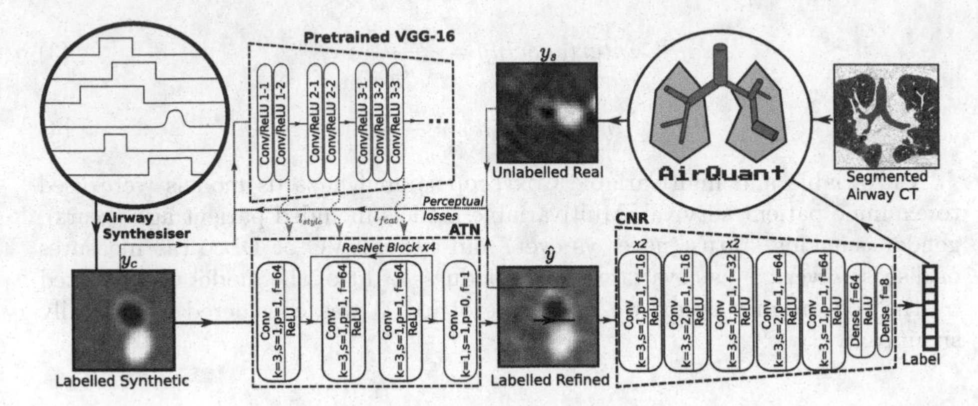

Fig. 1. Schematic demonstrating the data flows and model architectures. Also included is the architecture of the Airway Transfer Network (ATN) and Convolutional Neural Regressor (CNR). Where y_c, y_s and $\hat{y}$ refer to the notation used for calculating feature, l_{feat} and style, l_{style} losses from the particular activation layers of the pretrained VGG-16 model. AirQuant is an opensource airway analysis framework that can extract airway patches. The CNR model feeds measurements of the real airways back to AirQuant for final analysis.

factor in range of [0.0001, 0.1]. simGAN and ATN took 14 and 0.6 h respectively to converge during training. We qualitatively found that both simGAN and ATN produced refined images of optimal quality with a $\|l\|_1$ regularisation factor of 0.01.

Style-transfer from paintings to natural images show that larger-scale structure is transferred from the target image when training on losses of higher layers [11]. In order to maintain label correspondence between refiner input and output, we similarly only use the feature loss using the `relu3_3` activation layer. Style loss is computed from the two lower `relu1_2`, `relu2_2` activation layers only[3]. Figure 2 demonstrates qualitative results of our airway refinement method.

The CNR was trained with batch size in the interval [256, 2000] and learning rate of 0.001. Batch size of 2000 was chosen for its speed, and converged at around 40 epochs within one hour. The CNR achieves comparable results on ATN and simGAN refined images.

Figure 3 demonstrates qualitative results of our ATN method on real CT data. Table 1 shows results of the Cox regression survival analyses. The CNR when regressing to an airway feature demonstrated a strong association with mortality. This was despite the CNR label not perfectly aligning to the exact airway boundary.

[3] higher activation layers are considered in the supplementary material.

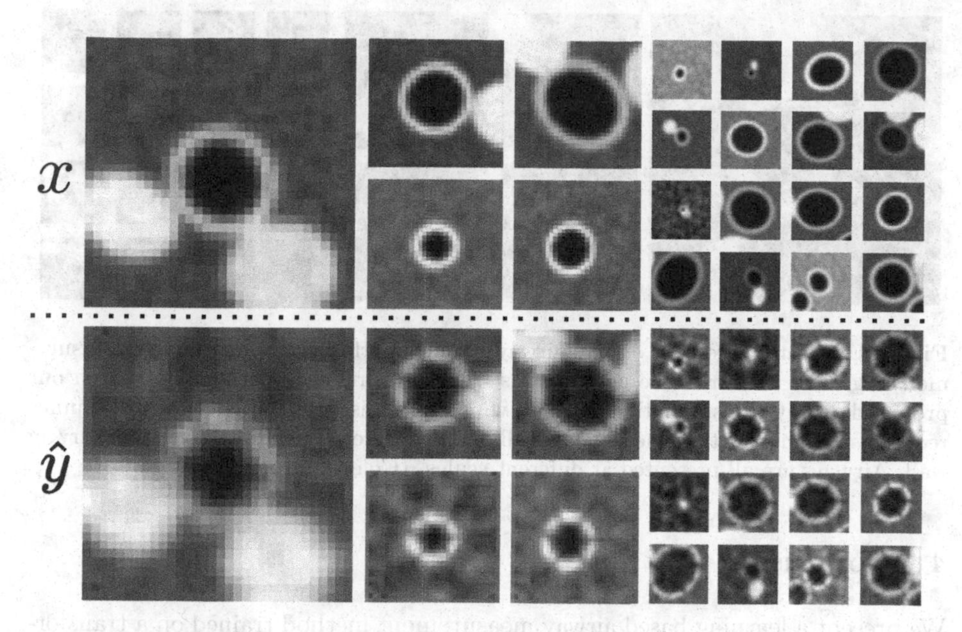

Fig. 2. Uncurated set of synthetic images x and output $\hat{y}$ of our airway transformation network in the same relative position below. Our model was trained to minimise perceptual losses. Airways are all represented at different scales.

Table 1. Cox proportional hazards results comparing mortality prediction of airway biomarkers derived by different measurement methods.

Method	Univariable (n = 113)		DLCo (n = 103)		FVC (n = 111)	
	C index	p-value	C index	p-value	C index	p-value
Volume						
FWHMesl [20]	0.61	0.00190	0.67	0.03031	0.68	0.03965
simGAN [16]	0.65	0.00006	0.68	0.00233	0.70	0.00086
Synthetic (no refinement)	0.61	0.00286	0.67	0.01040	0.69	0.00421
ATN (ours)	**0.67**	**0.00001**	**0.69**	**0.00013**	**0.71**	**<0.00001**
Intertapering						
FWHMesl [20]	0.55	0.07009	0.66	0.14999	0.68	0.08744
simGAN [16]	0.60	0.00925	0.67	0.03460	0.69	0.04764
Synthetic (no refinement)	0.60	0.00648	0.68	0.00692	0.69	0.00311
ATN (ours)	**0.62**	**0.00084**	**0.69**	**0.00062**	**0.70**	**0.00052**
Intratapering						
FWHMesl [20]	0.55	0.33623	0.66	0.93103	0.69	0.63837
simGAN [16]	0.59	0.09232	0.67	0.35460	0.69	0.48513
Synthetic (no refinement)	0.60	0.00537	**0.69**	0.00500	0.68	0.00263
ATN (ours)	**0.63**	**0.00026**	0.68	**0.00208**	0.69	**0.00192**

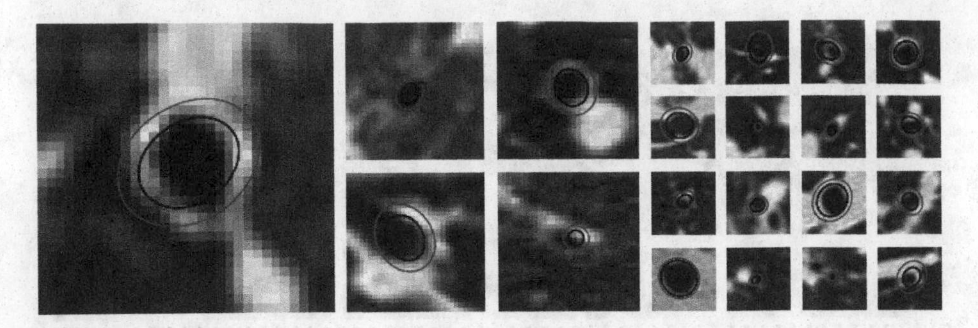

Fig. 3. Uncurated inference on real airway patches performed by our airway measurement regressor network. The network was trained on refined synthetic data from our proposed airway transformation network, which minimises perceptual losses. The inner red ellipse delineates the inner airway wall and the outer blue ellipse, the outer airway wall. Airways are all presented at different scales. (Color figure online)

4 Conclusion

We present a learning based airway measurement method trained on a transformation network that refines synthetic data using perceptual losses. Our model ATN was compared with a state-of-the-art model simGAN [16] and a physics based method FWHMesl. When assessing the clinical utility of ATN, we found that it was the strongest predictor of survival across all three airway biomarkers. We found that our method trains faster and with minimal complications, unlike a GAN framework. We expect future work to consider the generalisation of such a method, for example examining airways in patients with different diseases, images acquired on different scanner parameters and potentially on higher scale imaging such as micro-CT studies of the lungs.

Acknowledgements. This research was funded in whole or in part by the Wellcome Trust [209553/Z/17/Z]. For the purpose of open access, the author has applied a CC-BY public copyright licence to any author accepted manuscript version arising from this submission. AP is funded jointly by the Cystic Fibrosis Trust and EPSRC i4health, centre for doctoral training studentship. JJ was supported by a Wellcome Trust Clinical Research Career Development Fellowship and the NIHR UCLH Biomedical Research Centre, UK.

References

1. Biewald, L.: Experiment tracking with weights and biases (2020). https://www.wandb.com/, software available from wandb.com
2. Deng, J., Dong, W., Socher, R., Li, L.J., Li, K., Fei-Fei, L.: ImageNet: a large-scale hierarchical image database. In: 2009 IEEE Conference on Computer Vision and Pattern Recognition, pp. 248–255, June 2009. https://doi.org/10.1109/CVPR.2009.5206848

3. Estépar, R.S.J., Washko, G.G., Silverman, E.K., Reilly, J.J., Kikinis, R., Westin, C.-F.: Accurate airway wall estimation using phase congruency. In: Larsen, R., Nielsen, M., Sporring, J. (eds.) MICCAI 2006. LNCS, vol. 4191, pp. 125–134. Springer, Heidelberg (2006). https://doi.org/10.1007/11866763_16
4. Flaherty, K.R., et al.: Idiopathic pulmonary fibrosis. Am. J. Respir. Crit. Care Med. **174**(7), 803–809 (2006). https://doi.org/10.1164/rccm.200604-488OC
5. Gatys, L.A., Ecker, A.S., Bethge, M.: A neural algorithm of artistic style (2015). https://doi.org/10.48550/ARXIV.1508.06576, https://arxiv.org/abs/1508.06576
6. Goodfellow, I.J., et al.: Generative adversarial networks. arXiv:1406.2661 [cs, stat] (June 2014). http://arxiv.org/abs/1406.2661
7. Gu, S., et al.: Computerized identification of airway wall in CT examinations using a 3D active surface evolution approach. Med. Image Anal. **17**(3), 283–296 (2013). https://doi.org/10.1016/j.media.2012.11.003
8. Harrell Jr., F.E., Lee, K.L., Mark, D.B.: Multivariable prognostic models: issues in developing models, evaluating assumptions and adequacy, and measuring and reducing errors. Stat. Med. **15**(4), 361–387 (1996). https://doi.org/10.1002/(SICI)1097-0258(19960229)15:4⟨361::AID-SIM168⟩3.0.CO;2-4
9. He, K., Zhang, X., Ren, S., Sun, J.: Deep residual learning for image recognition, December 2015. https://doi.org/10.48550/arXiv.1512.03385
10. Hofmanninger, J., Prayer, F., Pan, J., Röhrich, S., Prosch, H., Langs, G.: Automatic lung segmentation in routine imaging is primarily a data diversity problem, not a methodology problem. Eur. Radiol. Exp. **4**(1), 1–13 (2020). https://doi.org/10.1186/s41747-020-00173-2
11. Johnson, J., Alahi, A., Fei-Fei, L.: Perceptual losses for real-time style transfer and super-resolution, March 2016. https://doi.org/10.48550/arXiv.1603.08155
12. Kiraly, A.P., Reinhardt, J.M., Hoffman, E.A., McLennan, G., Higgins, W.E.: Virtual bronchoscopy for quantitative airway analysis. In: Amini, A.A., Manduca, A. (eds.) Medical Imaging 2005: Physiology, Function, and Structure from Medical Images, vol. 5746, p. 369. International Society for Optics and Photonics, April 2005. https://doi.org/10.1117/12.595283
13. Kluvanec, D., Phillips, T.B., McCaffrey, K.J.W., Moubayed, N.A.: Using orientation to distinguish overlapping chromosomes, March 2022. https://doi.org/10.48550/arXiv.2203.13004
14. Kuo, W., Perez-Rovira, A., Tiddens, H., de Bruijne, M.: Airway tapering: an objective image biomarker for bronchiectasis. Eur. Radiol. **30**(5), 2703–2711 (2019). https://doi.org/10.1007/s00330-019-06606-w
15. Lederer, D.J., Martinez, F.J.: Idiopathic pulmonary fibrosis. N. Engl. J. Med. **378**(19), 1811–1823 (2018). https://doi.org/10.1056/NEJMra1705751
16. Nardelli, P., Ross, J.C., San José Estépar, R.: Generative-based airway and vessel morphology quantification on chest CT images. Med. Image Anal. **63**, 101691 (2020). https://doi.org/10.1016/j.media.2020.101691
17. Pakzad, A., et al.: Evaluation of automated airway morphological quantification for assessing fibrosing lung disease. Technical report, November 2021. arXiv:2111.10443, arXiv. https://doi.org/10.48550/ARXIV.2111.10443
18. Pakzad, A., Jacob, J.: Radiology of bronchiectasis. Clin. Chest Med. **43**(1), 47–60 (2022). https://doi.org/10.1016/j.ccm.2021.11.004
19. Paszke, A., et al.: Pytorch: an imperative style, high-performance deep learning library. In: Wallach, H., Larochelle, H., Beygelzimer, A., d' Alché-Buc, F., Fox, E., Garnett, R. (eds.) Advances in Neural Information Processing Systems, vol. 32, pp. 8024–8035. Curran Associates, Inc. (2019). http://papers.neurips.cc/paper/9015-pytorch-an-imperative-style-high-performance-deep-learning-library.pdf

20. Quan, K., et al.: Tapering analysis of airways with bronchiectasis. In: Angelini, E.D., Landman, B.A. (eds.) Medical Imaging 2018: Image Processing, vol. 10574, p. 87. SPIE, March 2018. https://doi.org/10.1117/12.2292306
21. Shrivastava, A., Pfister, T., Tuzel, O., Susskind, J., Wang, W., Webb, R.: Learning from simulated and unsupervised images through adversarial training (2017). https://doi.org/10.48550/ARXIV.1612.07828, https://arxiv.org/abs/1612.07828
22. Simonyan, K., Zisserman, A.: Very deep convolutional networks for large-scale image recognition. arXiv:1409.1556 [cs] (April 2015). https://doi.org/10.48550/ARXIV.1409.1556
23. Weibel, E.R.: Morphometry of the Human Lung. Springer, Berlin, Heidelberg (1963). https://doi.org/10.1007/978-3-642-87553-3_6
24. Willemink, M.J., et al.: Preparing medical imaging data for machine learning. Radiology **295**(1), 4–15 (2020). https://doi.org/10.1148/radiol.2020192224
25. Xie, M., Liu, X., Cao, X., Guo, M., Li, X.: Trends in prevalence and incidence of chronic respiratory diseases from 1990 to 2017. Respir. Res. **21**(1), 1–13 (2020). https://doi.org/10.1186/s12931-020-1291-8
26. Xu, Z., Bagci, U., Foster, B., Mansoor, A., Udupa, J.K., Mollura, D.J.: A hybrid method for airway segmentation and automated measurement of bronchial wall thickness on CT. Med. Image Anal. **24**(1), 1–17 (2015). https://doi.org/10.1016/j.media.2015.05.003
27. Yu, F., Koltun, V.: Multi-scale context aggregation by dilated convolutions. arXiv:1511.07122 [cs] (April 2016). https://doi.org/10.48550/ARXIV.1511.07122

Brain Imaging Generation with Latent Diffusion Models

Walter H. L. Pinaya[1(✉)], Petru-Daniel Tudosiu[1], Jessica Dafflon[2,3], Pedro F. Da Costa[4,5], Virginia Fernandez[1], Parashkev Nachev[6], Sebastien Ourselin[1], and M. Jorge Cardoso[1]

[1] Department of Biomedical Engineering, School of Biomedical Engineering and Imaging Sciences, King's College London, London, UK
walter.diaz_sanz@kcl.ac.uk

[2] Data Science and Sharing Team, Functional Magnetic Resonance Imaging Facility, National Institute of Mental Health, Bethesda, MD 20892, USA

[3] Machine Learning Team, Functional Magnetic Resonance Imaging Facility, National Institute of Mental Health, Bethesda, MD 20892, USA

[4] Psychology and Neuroscience, Institute of Psychiatry, King's College London, London, UK

[5] Centre for Brain and Cognitive Development, Birkbeck College, London, UK

[6] Institute of Neurology, University College London, London, UK

Abstract. Deep neural networks have brought remarkable breakthroughs in medical image analysis. However, due to their data-hungry nature, the modest dataset sizes in medical imaging projects might be hindering their full potential. Generating synthetic data provides a promising alternative, allowing to complement training datasets and conducting medical image research at a larger scale. Diffusion models recently have caught the attention of the computer vision community by producing photorealistic synthetic images. In this study, we explore using Latent Diffusion Models to generate synthetic images from high-resolution 3D brain images. We used T1w MRI images from the UK Biobank dataset (N = 31,740) to train our models to learn about the probabilistic distribution of brain images, conditioned on covariates, such as age, sex, and brain structure volumes. We found that our models created realistic data, and we could use the conditioning variables to control the data generation effectively. Besides that, we created a synthetic dataset with 100,000 brain images and made it openly available to the scientific community.

Keywords: Synthetic data · Diffusion models · Generative models · Brain imaging

---

W. H. L. Pinaya and P.-D. Tudosiu—Equal contribution.

Supplementary Information The online version contains supplementary material available at https://doi.org/10.1007/978-3-031-18576-2_12.

A. Mukhopadhyay et al. (Eds.): DGM4MICCAI 2022, LNCS 13609, pp. 117–126, 2022.
https://doi.org/10.1007/978-3-031-18576-2_12

1 Introduction

Deep neural networks fuelled several ground-breaking advancements in areas such as natural language processing and computer vision, where part of these improvements was attributed to the large amount of rich data used to train these networks, with some public datasets reaching millions of images and text sentences [8,26]. During the same period, medical image analysis also made remarkable breakthroughs by applying deep neural networks to solve tasks such as segmentation, structure detection, and computer-aided diagnosis (detailed review available at [20,27]). However, one current limitation of medical imaging projects is the lack of availability of large datasets. Medical data are costly and laborious to collect, and privacy concerns create challenges to data sharing by restricting publicly available medical datasets to up to a few thousand examples. This limitation creates a bottleneck on models' generalizability and hampers the rate at which cutting-edge methods are deployed in the clinical routine.

Generating synthetic data with privacy guarantees provides a promising alternative, allowing meaningful research to be carried out at scale [14,15,33]. Together with traditional data augmentation techniques (e.g., geometric transformations), these synthetic data could complement real data to dramatically increase the training set of machine learning models. Generative models learn the probability density function underlying the data they are trained on and can create realistic representations of examples which are different from the ones present in the training data by sampling from the learned distribution. However, generating meaningful synthetic data is not easy, especially when considering complex organs like the brain.

Nowadays, Generative Adversarial Networks (GANs) have been applied in various fields to create synthetic images, producing realistic and clear images and achieving impressive performance [7,32]. In the medical field, for example, [17] combine variational autoencoders with GANs to generate various modalities of whole brain volumes from a small training set and achieved a better performance compared to several baselines. However, since their study resized the images to a small volume before training, with a size of $64 \times 64 \times 64$ voxels, their synthetic medical images did not replicate many essential finer details. In addition, due to the prevalence of 3D high-resolution data in the field, researchers tend to have their models restrained by the amount of GPU memory available. To mitigate this problem, [31] proposed a 3D GAN with a hierarchical structure which is able to generate a low-resolution version of the image and anchor the generation of high-resolution sub-volumes on it. With this approach, the authors were able to generate impressive realistic 3D thorax CT and brain MRI with resolutions up to $256 \times 256 \times 256$ voxels. Despite generating great interest, GANs still come with inherent challenges, such as being notoriously unstable during training and failing to converge or to capture the variability of the generated data due to mode collapse issues [16].

Recently, diffusion models caught the attention of the machine learning community by showing promising results when synthesizing natural images. They have rivalled GANs in sample quality [9] while building upon a solid theoretical

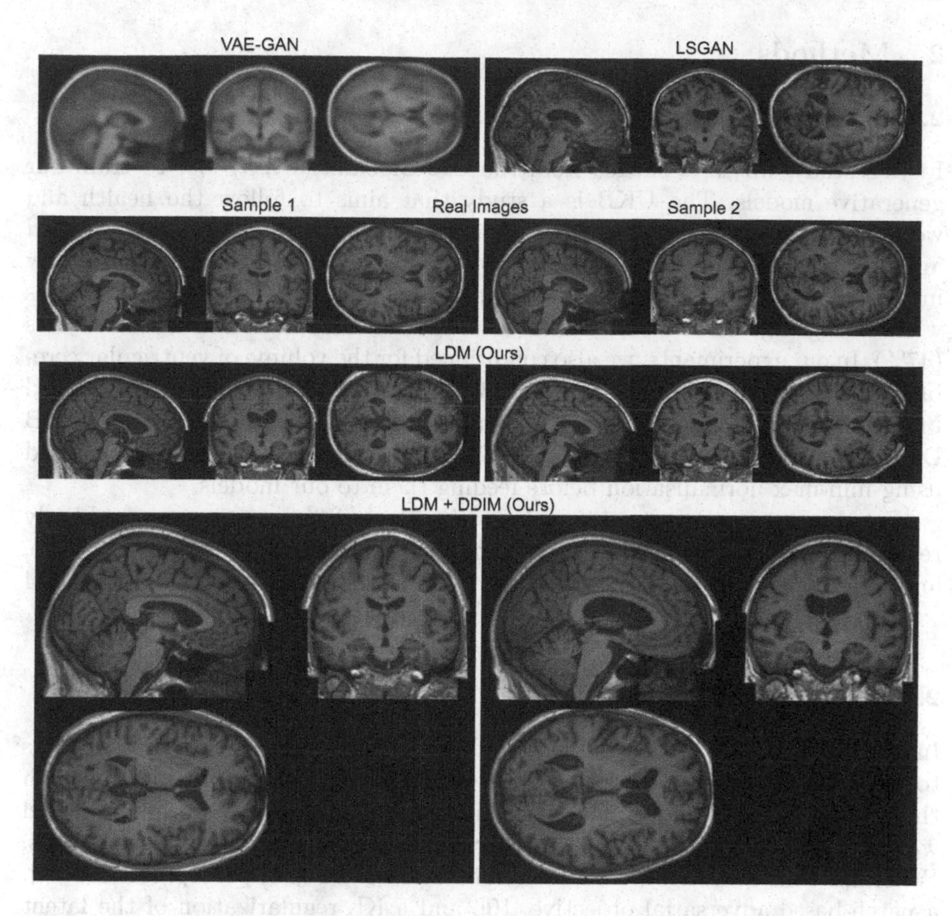

Fig. 1. Real and synthetic samples of head MRI generated using VAE-GAN, LSGAN, LDM and LDM+DDIM.

foundation. Not only have they reported impressive photorealism unconditioned images, but they have also been used to create images conditioned in classes and text sentences (using techniques like classifier-free guidance [13]), with exceptional results on models like Latent Diffusion Models [23], DALLE 2 [22], and Imagen [25].

In this study, we used diffusion models to create synthetic MRI images of the adult human brain. For that, we used 31,740 training images from the UK Biobank [30]to train our models. In order to efficiently scale the application of diffusion models to these high-resolution 3D data, we combined our diffusion models with compression models following the architecture of Latent Diffusion Models (LDM) [23]. Furthermore, we conditioned the image generations on age, gender, ventricular volume, and brain volume relative to the intracranial volume in order to generate realistic examples of brain scans with specific covariate values. We compared our synthetic images to state-of-the-art methods based on GANs, and we made our synthetic dataset comprising 100,000 brain images publicly available to the scientific community.

2 Methods

2.1 Datasets and Image Preprocessing

In this study, we used images from the UK Biobank (UKB) [30] to train our generative models. The UKB is a study that aims to follow the health and well-being of volunteer participants across the United Kingdom. Here, we used an early release of the project's data comprising 31,740 participants with T1w images. The dataset consists of healthy individuals aged between 44 and 82 years with average age of 63.6 ± 7.5 years (average $\pm$ SD) and 14,942 male subjects (47%). In our experiments, we also conditioned for the volume of ventricular cerebrospinal fluid (min-max: 6995.68 - 171375.0 mm^3; UKB Data-Field 25004) and brain volume normalised for head size (min-max: 1144240 - 1793910 mm^3; UKB Data-Field 25009). All variables used for model conditioning were normalised using min-max normalisation before feeding them to our models.

For the image pre-processing, we used UniRes[1] [2,3] to perform a rigid body registration to a common MNI space. The final images had 1 mm^3 as voxel size, and we cropped the image to obtain a volume of the head measuring $160 \times 224 \times 160$ voxels.

2.2 Generative Models

In our experiments, we used LDMs, which combine the use of autoencoders to compress the input data into a lower-dimensional latent representation with the generative modelling properties of diffusion models. The compression model was an essential step to allow us to scale to high-resolution medical images. We trained the autoencoder with a combination of L1 loss, perceptual loss [34], a patch-based adversarial objective [10], and a KL regularization of the latent space. The encoder maps the brain image to a latent representation with a size of $20 \times 28 \times 20$. After training the compression model, the latent representations of the training set are used as input to the diffusion model. Diffusion models [12,28] are generative models that convert Gaussian noise into samples from a learned data distribution via an iterative denoising process. Given a latent representation of an example from our training set, the diffusion process gradually destroys the structure of the data via a fixed Markov chain over 1000 steps by adding Gaussian noise using a fixed linear variance schedule. The reverse process is also modelled as a Markov chain which learns to recover the original input from the noisy one. We conditioned our models according to age, gender, ventricular volume, and brain volume relative to the intracranial volume. To perform this conditioning, we used a hybrid approach combining the concatenation of the conditioning values with the inputted latent representation (i.e., as additional channels) and the use of cross-attention mechanisms, as proposed in [23]. Training and model details are available in the supplementary material.

[1] https://github.com/brudfors/UniRes.

3 Experiments

3.1 Sampling Quality

Figure 1 shows images generated using LDMs compared to real images and the baselines (i.e., VAE-GAN [18] and LSGAN [21]). Unlike the baselines, we observe that the LDMs were able to sample high-quality images with sharp details and realistic textures. Besides that, training the diffusion models at such a high resolution was much more stable and easier to achieve convergence when compared to the GAN-based baselines. The baselines required a meticulous design of the interaction between discriminator and generator, and they presented problems of mode collapse, showcasing the problems of GAN-based applied in such high-resolution 3D images. Therefore, we will refine and expand our comparisons with other baselines in future works.

We also obtained quantitative metrics about the performance of our models. We used the Fréchet Inception Distance (FID) [11] to measure how realistic the synthetic images are. A small FID indicates that the distribution of the generated images is similar to the distribution of the real images. The FID was calculated using an approach similar to [31], where features were extracted using a pre-trained Med3D [5]. We also measured the generation diversity with the Multi-Scale Structural Similarity Metric (MS-SSIM) and 4-G-R-SSIM [4,19,24], where a value close to 0 suggests high diversity. Here, we presented the MS-SSIM for comparison with previous studies, but we also added the 4-G-R-SSIM as it has been shown to have better image quality assessment. We compute the average values from 1000 sample pairs. Table 1 shows the quantitative results for different models used for the image synthesis.

Table 1. Quantitative evaluation of the synthetic images on the UK Biobank. We used the Fréchet Inception Distance (FID) to verify how realistic are the images, and we used the multi-scale structural similarity metric (MS-SSIM) and 4-G-R-SSIM to evaluate the generation diversity. We used 50 timesteps when sampling our models with DDIM sampler. Real images measures were obtained comparing the training set with 1000 brain image from a hold-out test set.

	FID ↓	MS-SSIM ↓	4-G-R-SSIM ↓
LSGAN	0.0231	0.9997	0.9969
VAE-GAN	0.1576	0.9671	0.8719
LDM	**0.0076**	**0.6555**	**0.3883**
LDM + DDIM	0.0080	0.6704	0.3957
Real images	0.0005	0.6536	0.3909

Recently, different methods have been proposed to speed up the reverse process (e.g., Denoising Diffusion Implicit Models - DDIM), reducing by 10× 50× the number of necessary reverse steps [29]. Using the DDIM sampler, we reduced the number of timesteps from 1000 steps to only 50. This improved our sampling time from an average of 142.3 ± 1.6 s per sample to 7.6 ± 0.2 s per sample @ NVIDIA TITAN RTX with minimum loss in performance (Table 1). Because of this boost in processing time and a minimal performance loss, we are using the LDM with the DDIM sampler for all the remaining analyses.

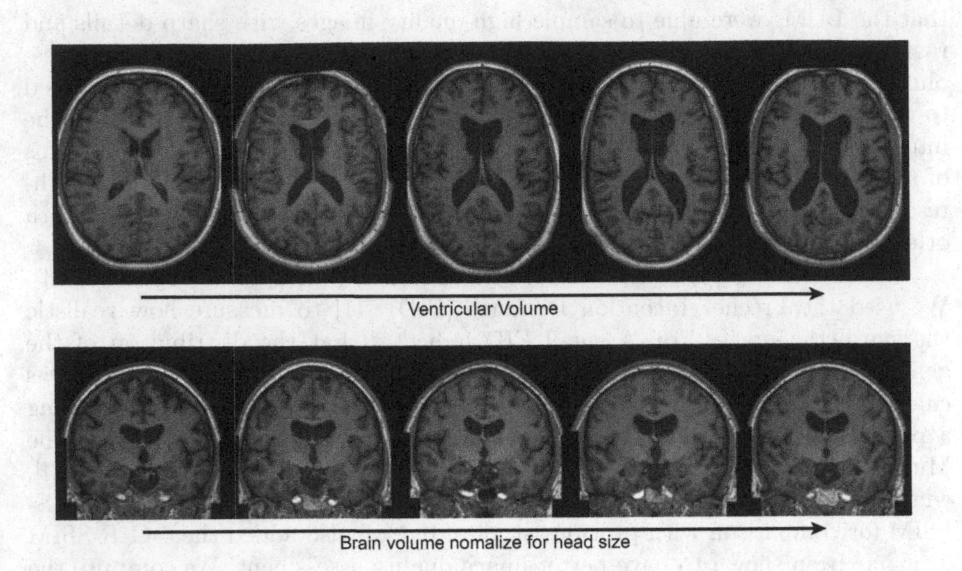

Fig. 2. Conditioned sampling varying the ventricular volume and the brain volume normalised by the intracranial volume. In both rows, we kept the other variables constant.

3.2 Conditioning Evaluation

Using the hybrid conditioning approach [23], we were able to condition our models and generate brain images where we can specify the age, sex, ventricular volume, and brain volume. As we can observe in Fig. 2, our model was able to learn representations conditioned on regional (i.e., ventricular volume) and global (i.e., brain volume) volumes.

In order to quantitatively evaluate the conditioning, we used SynthSeg[2] [1] to measure the volumes of the ventricles of 1000 synthetic brains. In this analysis, we measured the combination of the left and right lateral ventricles and the left and right inferior lateral ventricles. We then computed the Pearson correlation between the obtained volumes and the inputted conditioning values. Using this approach, we observed a high correlation coefficient of 0.972, which demonstrates the effectiveness of conditioning on our model (Fig. 3).

[2] https://github.com/BBillot/SynthSeg.

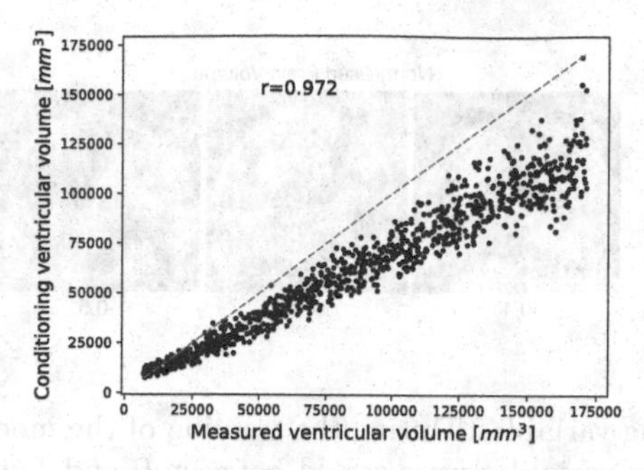

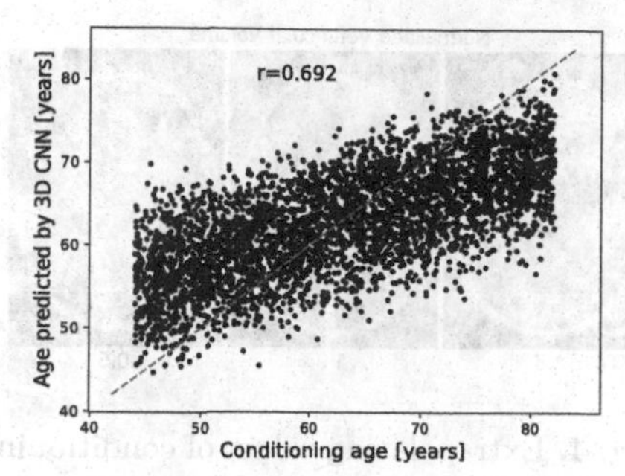

Fig. 3. Conditioning analysis. Left) Correlation between conditioning ventricular volume vs ventricular measured with SynthSeg. Right) Correlation between conditioning age vs brain age predicted by 3D convolutional neural network (3D CNN).

Additionally, we verified how well the effectiveness of conditioning brain generation by age. To this end, we used discriminative models to perform the task of brain age prediction, where we predict chronological age based on the brain image. In our study, we used the 3D convolution neural network proposed in [6], trained on the same training set used in the LDM training. After training the model, we verify how well the predicted age approximated the inputted age of the synthetic dataset. As shown in Fig. 3, our model presented a high correlation between the inputted conditioning and the predicted age ($r = 0.692$).

Finally, we verified how our model extrapolates the conditioning variables for values never shown during training. Figure 4 presents samples where we used a normalised ventricular value higher than 1; in this case, we can see abnormally huge ventricles when using values of 1.5 and 1.9. If we use a negative value (e.g., -0.5), an image without ventricles is generated. Similarly, if we use negative values for the brain normalised for head size, the brain exhibits signs of neurodegeneration, showing smaller volumes of white and grey matter. These findings suggest that our models learned the concepts behind these conditioning variables during training.

3.3 Synthetic Dataset

We made a synthetic dataset of 100,000 human brain images generated by our model openly available to the community. This dataset is available at Academics Torrents[3], FigShare[4], and HDRUK Gateway[5], together with the conditioning information.

[3] https://academictorrents.com/details/63aeb864bbe2115ded0aa0d7d36334c026f0660b.

[4] https://figshare.com/.

[5] https://www.healthdatagateway.org/.

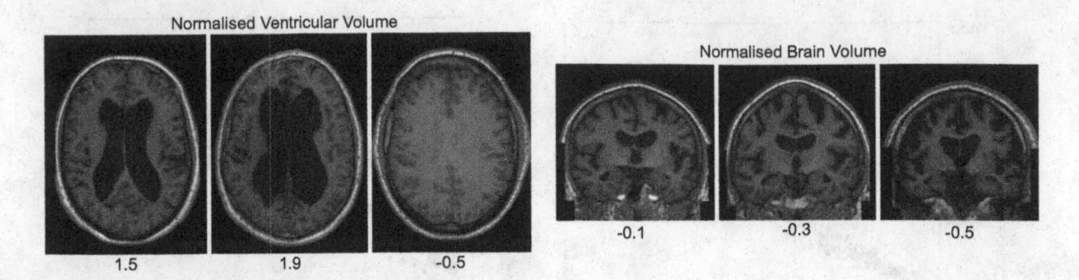

Fig. 4. Extrapolating values of conditioning variables. During the training of the models, the inputted values of the conditioning variables were scaled between 0 and 1. In this experiment, we tried values outside of this range, and we observed that our model could extrapolate the representation of brain and ventricular volumes, showing that it learned the concept of these variables.

4 Conclusions

In our study, we were able to train diffusion models to effectively generate synthetic brain images that replicate properties from the training images. As it is the case with natural image generation, our diffusion models outperform alternative GANs-based methods in an unconditioned scenario. Additionally, we demonstrated how our methods could be conditioned on covariates such as age, sex, and brain structure volumes to produce the expected representation. In future works, we will develop models that use other scanning modalities as conditioning, such as images and radiological reports. By making the synthetic dataset openly available, this work also addresses one of the biggest limitations in medical machine learning - the challenge of obtaining large imaging datasets - while not posing threats to privacy infringements. In sum, our results show that LDMs are promising models to be explored in medical image generation.

Acknowledgements. WHLP and MJC are supported by Wellcome Innovations [WT213038/Z/18/Z]. PTD is supported by the EPSRC Research Council, part of the EPSRC DTP, grant Ref: [EP/R513064/1]. JD is supported by the Intramural Research Program of the NIMH (ZIC-MH002960 and ZIC-MH002968). PFDC is supported by the European Union's HORIZON 2020 Research and Innovation Programme under the Marie Sklodowska-Curie Grant Agreement No 814302. PN is supported by Wellcome Innovations [WT213038/Z/18/Z] and the UCLH NIHR Biomedical Research Centre. This research has been conducted using the UK Biobank Resource (Project number: 58292).

References

1. Billot, B., et al.: Synthseg: domain randomisation for segmentation of brain MRI scans of any contrast and resolution. arXiv preprint arXiv:2107.09559 (2021)
2. Brudfors, M., Balbastre, Y., Nachev, P., Ashburner, J.: MRI super-resolution using multi-channel total variation. In: Nixon, M., Mahmoodi, S., Zwiggelaar, R. (eds.) MIUA 2018. CCIS, vol. 894, pp. 217–228. Springer, Cham (2018). https://doi.org/10.1007/978-3-319-95921-4_21
3. Brudfors, M., Balbastre, Y., Nachev, P., Ashburner, J.: A tool for super-resolving multimodal clinical MRI. arXiv preprint arXiv:1909.01140 (2019)
4. Chen, G.H., Yang, C.L., Xie, S.L.: Gradient-based structural similarity for image quality assessment. In: 2006 International Conference on Image Processing, pp. 2929–2932. IEEE (2006)
5. Chen, S., Ma, K., Zheng, Y.: Med3d: transfer learning for 3d medical image analysis. arXiv preprint arXiv:1904.00625 (2019)
6. Cole, J.H., et al.: Predicting brain age with deep learning from raw imaging data results in a reliable and heritable biomarker. Neuroimage **163**, 115–124 (2017)
7. Creswell, A., White, T., Dumoulin, V., Arulkumaran, K., Sengupta, B., Bharath, A.A.: Generative adversarial networks: an overview. IEEE Signal Process. Mag. **35**(1), 53–65 (2018)
8. Deng, J., Dong, W., Socher, R., Li, L.J., Li, K., Fei-Fei, L.: Imagenet: a large-scale hierarchical image database. In: 2009 IEEE Conference on computer Vision and Pattern Recognition, pp. 248–255. IEEE (2009)
9. Dhariwal, P., Nichol, A.: Diffusion models beat gans on image synthesis. Adv. Neural. Inf. Process. Syst. **34**, 8780–8794 (2021)
10. Esser, P., Rombach, R., Ommer, B.: Taming transformers for high-resolution image synthesis. In: Proceedings of the IEEE/CVF Conference on Computer Vision and Pattern Recognition, pp. 12873–12883 (2021)
11. Heusel, M., Ramsauer, H., Unterthiner, T., Nessler, B., Hochreiter, S.: Gans trained by a two time-scale update rule converge to a local Nash equilibrium. Adv. Neural Inf. Process. Syst. **30** (2017)
12. Ho, J., Jain, A., Abbeel, P.: Denoising diffusion probabilistic models. Adv. Neural. Inf. Process. Syst. **33**, 6840–6851 (2020)
13. Ho, J., Salimans, T.: Classifier-free diffusion guidance. In: NeurIPS 2021 Workshop on Deep Generative Models and Downstream Applications (2021)
14. Jordon, J., et al.: Synthetic data-what, why and how? arXiv preprint arXiv:2205.03257 (2022)
15. Jordon, J., Wilson, A., van der Schaar, M.: Synthetic data: opening the data floodgates to enable faster, more directed development of machine learning methods. arXiv preprint arXiv:2012.04580 (2020)
16. Kodali, N., Abernethy, J., Hays, J., Kira, Z.: On convergence and stability of gans. arXiv preprint arXiv:1705.07215 (2017)
17. Kwon, G., Han, C., Kim, D.: Generation of 3D brain MRI using auto-encoding generative adversarial networks. In: Shen, D., et al. (eds.) MICCAI 2019. LNCS, vol. 11766, pp. 118–126. Springer, Cham (2019). https://doi.org/10.1007/978-3-030-32248-9_14
18. Larsen, A.B.L., Sønderby, S.K., Larochelle, H., Winther, O.: Autoencoding beyond pixels using a learned similarity metric. In: International Conference on Machine Learning, pp. 1558–1566. PMLR (2016)

19. Li, C., Bovik, A.C.: Content-partitioned structural similarity index for image quality assessment. Signal Process. Image Commun. **25**(7), 517–526 (2010)
20. Lundervold, A.S., Lundervold, A.: An overview of deep learning in medical imaging focusing on MRI. Z. Med. Phys. **29**(2), 102–127 (2019)
21. Mao, X., Li, Q., Xie, H., Lau, R.Y., Wang, Z., Paul Smolley, S.: Least squares generative adversarial networks. In: Proceedings of the IEEE International Conference on Computer Vision, pp. 2794–2802 (2017)
22. Ramesh, A., Dhariwal, P., Nichol, A., Chu, C., Chen, M.: Hierarchical text-conditional image generation with clip latents. arXiv preprint arXiv:2204.06125 (2022)
23. Rombach, R., Blattmann, A., Lorenz, D., Esser, P., Ommer, B.: High-resolution image synthesis with latent diffusion models. In: Proceedings of the IEEE/CVF Conference on Computer Vision and Pattern Recognition, pp. 10684–10695 (2022)
24. Rouse, D.M., Hemami, S.S.: Analyzing the role of visual structure in the recognition of natural image content with multi-scale SSIM. In: Human Vision and Electronic Imaging XIII, vol. 6806, pp. 410–423. SPIE (2008)
25. Saharia, C., et al.: Photorealistic text-to-image diffusion models with deep language understanding. arXiv preprint arXiv:2205.11487 (2022)
26. Schuhmann, C., et al.: Laion-400m: open dataset of clip-filtered 400 million image-text pairs. arXiv preprint arXiv:2111.02114 (2021)
27. Shen, D., Wu, G., Suk, H.I.: Deep learning in medical image analysis. Annu. Rev. Biomed. Eng. **19**, 221 (2017)
28. Sohl-Dickstein, J., Weiss, E., Maheswaranathan, N., Ganguli, S.: Deep unsupervised learning using nonequilibrium thermodynamics. In: International Conference on Machine Learning, pp. 2256–2265. PMLR (2015)
29. Song, J., Meng, C., Ermon, S.: Denoising diffusion implicit models. arXiv preprint arXiv:2010.02502 (2020)
30. Sudlow, C., et al.: Uk biobank: an open access resource for identifying the causes of a wide range of complex diseases of middle and old age. PLoS Med. **12**(3), e1001779 (2015)
31. Sun, L., Chen, J., Xu, Y., Gong, M., Yu, K., Batmanghelich, K.: Hierarchical amortized gan for 3D high resolution medical image synthesis. IEEE J. Biomed. Health Inform. **26**(8), 3966–3975 (2022)
32. Wang, L., Chen, W., Yang, W., Bi, F., Yu, F.R.: A state-of-the-art review on image synthesis with generative adversarial networks. IEEE Access **8**, 63514–63537 (2020)
33. Wang, T., et al.: A review on medical imaging synthesis using deep learning and its clinical applications. J. Appl. Clin. Med. Phys. **22**(1), 11–36 (2021)
34. Zhang, R., Isola, P., Efros, A.A., Shechtman, E., Wang, O.: The unreasonable effectiveness of deep features as a perceptual metric. In: Proceedings of the IEEE Conference on Computer Vision and Pattern Recognition, pp. 586–595 (2018)

Author Index